Study Guide for

Strong/DeVault/Sayad's
The Marriage and Family Experience
Intimate Relationships in a Changing Society
Seventh Edition

Carol Mertens
The University of Iowa

Wadsworth Publishing Company
I(T)P® An International Thomson Publishing Company

Belmont, CA • Albany, NY • Bonn • Boston • Cincinnati • Detroit • Johannesburg • London
Madrid • Melbourne • Mexico City • New York • Paris • Singapore • Tokyo • Toronto • Washington

COPYRIGHT © 1998 by Wadsworth Publishing Company
A Division of International Thomson Publishing Inc.
I(T)P® The ITP logo is a registered trademark under license.

Printed in the United States of America
1 2 3 4 5 6 7 8 9 10

For more information, contact Wadsworth Publishing Company, 10 Davis Drive, Belmont, CA 94002, or electronically at http://www.thomson.com/wadsworth.html

International Thomson Publishing Europe
Berkshire House 168-173
High Holborn
London, WC1V 7AA, England

International Thomson Editores
Campos Eliseos 385, Piso 7
Col. Polanco
11560 México D.F. México

Thomas Nelson Australia
102 Dodds Street
South Melbourne 3205
Victoria, Australia

International Thomson Publishing Asia
221 Henderson Road
#05-10 Henderson Building
Singapore 0315

Nelson Canada
1120 Birchmount Road
Scarborough, Ontario
Canada M1K 5G4

International Thomson Publishing Japan
Hirakawacho Kyowa Building, 3F
2-2-1- Hirakawacho
Chiyoda-ku, Tokyo 102, Japan

International Thomson Publishing GmbH
Königswinterer Strasse 418
53227 Bonn, Germany

International Thomson Publishing Southern Africa
Building 18, Constantia Park
240 Old Pretoria Road
Halfway House, 1685 South Africa

All rights reserved. No part of this work covered by the copyright hereon may be reproduced or used in any form or by any means—graphic, electronic, or mechanical, including photocopying, recording, taping, or information storage and retrieval systems—without the written permission of the publisher.

ISBN 0-534-53758-8

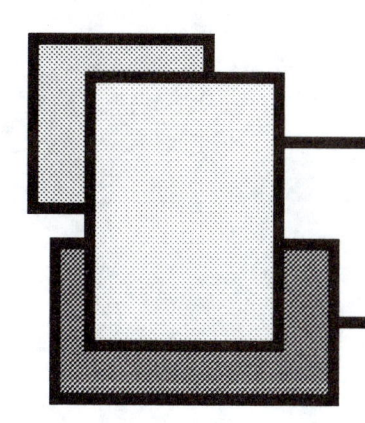

Contents

Chapter 1   The Meaning of Marriage and Family 1
Chapter 2   Studying Marriage and Family 19
Chapter 3   Contemporary Gender Roles 41
Chapter 4   Friendship, Love, and Commitment 57
Chapter 5   Communication and Conflict Resolution 71
Chapter 6   Pairing and Singlehood 87
Chapter 7   Understanding Sexuality 103
Chapter 8   Pregnancy and Childbirth 121
Chapter 9   Marriage as Process: Family Life Cycles 137
Chapter 10   Parents and Children 153
Chapter 11   Marriage, Work, and Economics 171
Chapter 12   Families and Wellness 193
Chapter 13   Family Violence and Sexual Abuse 209
Chapter 14   Coming Apart: Separation and Divorce 223
Chapter 15   New Beginnings: Single-Parent
             Families and Stepfamilies 239
Chapter 16   Marriage and Family Strengths 255

To Bryan

The impact you have had on the world of marriage and the family will always remain, even though you are no longer with us physically. Thank you for your gentle, caring way of connecting with others. Although your untimely death reminds us all of our own humanity, your courage and optimism remain as examples for us to follow. You were, and will always be, a good friend and a kindred spirit.

PREFACE

Welcome to the study guide which accompanies the Seventh Edition of *The Marriage and Family Experience*. Bryan Strong, Christine DeVault, and Barbara Sayad have written an excellent text which is scientifically sound, broad in scope, retains a quality of warmth and humor, and is relevant to the personal lives of its readers.

This study guide has been specifically designed to correspond with the textbook. To achieve the maximum use of this guide, it is recommended that you follow the suggestions below.

READING THE TEXT BOOK CHAPTER

First, examine the chapter of the text which has been assigned. Preview the material. Glance through the headings and note the key terms which are in italics or boldface type. Try to get a feeling for the material which this chapter covers. Do you see things which are of particular interest to you?

Examine the preview questions at the beginning of the chapter and attempt to answer them based on your present understanding of the issues. Look at photographs and charts. Do they relate to issues about which you are curious? Now examine the quotes in the margins. You will find that they come from a variety of sources, both classic and modern, because love, marriage, and relationships have been affecting human beings since the earliest times.

Go back to the beginning of the chapter, and read the material carefully, being sure to notice the terms which are emphasized. As you read, you may wish to underscore certain terms and key points for emphasis and later review, or make notes on certain points in the margins or in your notebook. The *Perspectives* have been added to provide you examples and increase your understanding of the material.

You may retain more of the material if you do not try to read an entire chapter at one time. Try to think critically about the material presented in each chapter. Does the material correspond with your previous understandings of marriage and the family? Were you surprised by some of the information? Did you change your opinions on anything? Did you disagree with the authors' conclusions? Did the authors seem to present a particular point of view?

Review the preview questions at the beginning of the chapter. Did you answer them correctly? Were you surprised by the answers? Look at the references, citations and quotes. How did the authors support their statements? What kind of research or writings did they cite? Did some ideas seem more strongly supported than others?

USING THE STUDY GUIDE

When you feel that you have a good understanding of the major ideas and concepts in the chapter, turn to the study guide. The first section is entitled **Main Focus**. This is a brief summary of key ideas presented in the chapter.

The second section, **Goals of this Chapter**, lists the main objectives of the chapter. If you have truly comprehended the chapter, you should be able to demonstrate mastery of each goal. If you have not mastered the goals, reread the material until you are able to meet the goal. You are not a failure if you have return to the book. The objective is full comprehension, not instant mastery. Therefore, you will probably benefit from revisiting the text and using it in coordination with the study guide.

The third section is a list of **Key Terms and Ideas** in the order which they appear in the text and study guide outline. It is important to understand and be able to define each of these terms. Examine each of the terms, ideas, and concepts, to see if you are clear about their meaning. If not, you might wish to reexamine the text or go to the next section for review, and come back to this section later.

The section entitled **Chapter Outline** is meant for review purposes only and is not intended to supplant the text. It provides an overview of the chapter, but does not always define terms, nor include all of the information and details of the chapter. The key terms and ideas are in boldface type so that you may see them in context. The terms are bolded if they appear in the section **Key Terms and Ideas**, and they are bolded only once, even though they may appear later.

The next section, **Test Your Comprehension**, involves a chart in which you must organize terms or concepts in a meaningful way. You may sometimes find this section difficult, but work with it, as it will often be useful for putting material in a perspective or framework which will enable you to grasp the major ideas more completely. You may have to turn to the book or your lecture materials to complete this section.

Each chapter contains a **Self-Discovery** section designed to help you relate the material to your own personal experience. There may not be a "wrong" or "right" response, only your personal perspectives or insights. Although this may make you a bit uncomfortable, material which you can relate to your own experience will be much more meaningful for you.

Each chapter has a section of **Self-Quizzes** composed of multiple choice and true/false questions. Occasionally, there will also be matching questions as well. Answer all of the questions, then check your answers with the **Key to Self Quizzes** at the end of the study guide chapter. Go back to the text if you need to.

The **Discuss Briefly** section includes short answer essays. This section is not always in the same order within each chapter, as the essays are frequently focused around a previous task.

Some of the chapters include a **Mini-Assignment**, which occasionally involves an activity outside of the classroom. Your instructor may ask you to do these activities. You may also find them beneficial and interesting to complete on your own. In many cases, these activities will relate the ideas and theories found in the text to your personal life experience.

Some of the chapters have a **Just for Fun** section which may include personal examinations of your ideas, values. There may also be humorous inserts. These activities help to provide some insight into who you are or how you see the world. Sometimes it is fun to do these exercises with a "significant other" or friend and compare your responses to theirs.

Each chapter always concludes with the **Key to Self-Quizzes** and a referral to text pages which offer **Suggested Readings**. While the **Key to Self-Quizzes** will tell you if your answers are correct, it is important to examine which answers were incorrect. It is suggested that you refer back to the chapter outline or the text to more fully understand incorrect responses and/or "good guesses."

CONCLUDING COMMENTS

I hope that you find this study guide helpful to you and personally relevant. Its usefulness will depend greatly on your type of learning skills and the amount of time that you spend with these materials.

The usefulness of study guides varies from student to student. The overall helpfulness of the study guide will depend on your personal style of learning and the amount of time you are willing to invest. Some instructors may require material from the study guide as part of the class assignment; other instructors may suggest that the student use all or part of these materials, as they prove helpful. A conscientious use of this study guide should significantly improve your mastery of the information provided in the text. I advise you to make the most of these materials to enhance your comprehension.

If you have to return to the textbook frequently, do not become frustrated. Remember, the goal is mastery of the material. You are not a failure if you have to check your answers or go back to review information… you are a smart and dedicated student!

I sincerely hope this material will provide new insights, knowledge, and wisdom to help you understand others while improving your own significant relationships.

Carol Mertens, CFLE
The University of Iowa

CHAPTER 1
The Meaning of Marriage and Family

MAIN FOCUS

Chapter One examines marriage, family, and television; definitions of marriage and family; functions of marriage and families; the concept of wellness; extended families and kinship; and an historical perspective of marriage and the family.

GOALS OF THIS CHAPTER

To demonstrate mastery of this chapter, you should be able to:
1. Describe how various American family forms are portrayed on American television.
2. Distinguish television stereotypes from the research based perception of American families.
3. Define marriage and its various forms throughout the world.
4. Explain why defining family is more complex than defining marriage.
5. List and describe the functions of marriages and families.
6. Define wellness, it's components, and how they interact to influence marriage.
7. Explain the advantages to living in families.
8. Define and describe the kinship system, including conjugal, consanguineous, and affiliated kinship relationships.
9. Provide an historical perspective of marriage and the family.
10. Explain the fundamental changes in American marriages and families.

Chapter 1

KEY TERMS AND IDEAS

The following terms, ideas, and concepts are listed in the order in which they appear in Chapter One. Be sure that you understand and can define each of the following:

cultivation theory	compadres	economies of scale
marriage	ie	kinship system
monogamy	clan	conjugal relationships
polygamy	nuclear family	consanguineous relationships
polygyny	traditional family	spirit marriage
polyandry	wellness	patrilineal
serial monogamy or modified polygamy	concubine	matrilineal
	couvade	patriarchy
family	family of orientation (origin)	bundling
affiliated kin	family of cohabitation	companiate marriages
kin	family of procreation	

CHAPTER ONE OUTLINE

I. INTRODUCTION TO THE CHAPTER
   A. Many of our own ideas about families are not formed by real-world experiences, but by television experiences.
   B. Television has had a significant influence in shaping our ideas, values, and beliefs about marriage, family, and other relationships.

II. MARRIAGE, FAMILY, AND TELEVISION
   A. Television may well be the most pervasive medium in our culture.
      1. Television transmits and reinforces social values.
      2. Television provides models which influence how we interact with others and how we expect others to interact with us.
      3. **Cultivation theory** asserts that there are consistent images, themes, and stereotypes which cut across programming genres to form a more or less consistent world view.
      4. Heavy TV viewers tend to have different beliefs and attitudes about the world than light TV viewers.
   B. Until the mid-1980s, the ideal family depicted on television was the nuclear family consisting of husband, wife, and their dependent children.
      1. In the mid-1980s, the primacy of the nuclear family on television was challenged by single-parent families, all-female families, all-male families, and multiracial families.
      2. Regardless of family form, certain themes run through sitcoms: love overcomes all adversity; divorce is rare; conflict is easily resolved through manipulation; problems are solved through humor; physical appearance is especially important.

C. Television portrays working-class and middle-class families differently.
D. Single-parent families and stepfamilies on television are usually formed as a result of a spouse's death rather than a divorce or birth to an unmarried woman.
   1. There are nearly seven times as many male-headed single-parent families on television than there are in reality.
   2. Single mothers are portrayed as looking for partners, while single fathers are represented as being satisfied with their single state.
   3. In reality, family dynamics are significantly different among single-parent families, step-families, and intact families; on TV they are portrayed as remarkably similar.
E. Most family interactions in sitcoms reflect marital stereotypes.
   1. Sex portrayed on television is usually premarital or extramarital.
   2. In reality, most sexual interactions take place within marriage.
F. Whites are overrepresented in television: African Americans, Latinos, Asian Americans, and Native Americans are underrepresented.
G. While the most offensive stereotypes have been eliminated from contemporary television, they endure in late-night reruns and syndication.
   1. Stereotypes abound in television families, however, many sitcoms value love and family relationships.
   2. In reality, the world is much more complicated than it is portrayed in sitcom families.
   3. The realities of daily life impinge on our intimate relationships.

III. WHAT IS MARRIAGE? WHAT IS FAMILY?
A. **Marriage** is a legally recognized union between a man and a woman in which they are united sexually, cooperate economically, and may give birth to, adopt, or rear children.
   1. Marriage differs among cultures and has changed historically in our society.
   2. The belief that marriage is divinely instituted is common to many religions, including many tribal religions, throughout the world.
   3. Each state enacts its own laws regulating marriage.
   4. **Monogamy**, one husband and one wife, is the only legal form of marriage in Western cultures: Worldwide, it is practiced by only 24 percent of known cultures.
   5. The preferred marital arrangement worldwide is **polygamy**, having more than one wife or husband.
      a. Eighty-four percent of the world's cultures study (still a minority of the world's population) practice or accept **polygyny**, having two or more wives: **Polyandry**, having two or more husbands, is quite rare.
      b. Within polygamous societies plural marriages are a minority, a sign of status relatively few people can afford.
      c. Considering the high divorce rate in America, the marriage system might be called **serial monogamy** or **modified polygamy**.

B. Defining **family** is even more complex than defining marriage.
   1. While most family members (**kin**) are related by descent, marriage, remarriage, or adoption, some are **affiliated kin** (unrelated individuals who feel like and are treated like relatives).
   2. Among Latinos, **compadres** (godparents) are considered family members.
   3. Among some Japanese Americans, the **ie** (living members of the extended family as well as deceased and yet-to-be-born family members) is the traditional family.
   4. Among many traditional Native American tribes, the **clan** (a group of related families) is regarded as the fundamental family unit.
   5. A major reason for difficulty in defining **family** is that we tend to think that the "real" family is the nuclear family or the **traditional family**.
      a. The **nuclear family** (a family consisting of mother, father, and children) is merely an idea or model we have about families.
      b. The traditional family is the middle-class nuclear family in which the women's primary roles are wife and mother and the men's primary roles are husband and breadwinner.
      c. The traditional family is the nuclear family wrapped in nostalgia and inequality; it exists more in the imagination than it ever did in reality.

IV. FUNCTIONS OF MARRIAGES AND FAMILIES
   A. Regardless of form, the family generally performs four important functions: (1) it provides a source of intimate relationships; (2) it acts as a unit of economic cooperation and consumption; (3) it may produce and socialize children; and (4) it assigns social roles and status to individuals.
   B. Intimacy is a primary human need.
      1. Human companionship strongly influences health.
      2. The need for intimacy, whether satisfactory or not, may keep unhappy marriages together indefinitely; loneliness may be a terrible specter.
      3. The need for intimacy is so powerful that we may even rely upon pets if intimacy needs are not met by humans.
   C. Economic cooperation, which traditionally divides labor along gender lines, varies from culture to culture.
      1. Among the Namibikwara, fathers care for babies; the chief's **concubines**, (secondary wives in polygamous societies) prefer hunting over domestic activities.
      2. Only a woman's ability to give birth and produce milk is biologically determined; some cultures practice **couvade** (ritualized childbirth in which the male gives birth to the child's spirit when his partner gives physical birth).
      3. The family is commonly thought of as a consuming unit, but it also continues to be an important producing unit.

D. The family makes society possible by producing (or adopting) and rearing children to replace older members of society who have died.
   1. Technological change has affected reproduction.
   2. The family traditionally has been responsible for socialization of children; this function is dramatically shifting away from the family.
E. We fulfill various social roles as family members and these roles provide us with much of our identity.
   1. The **family of orientation (origin)** is the family one grows up in.
   2. The **family of cohabitation** (traditionally known as **family of procreation**) is the family we form through living or cohabiting with another person.
   3. The status we are given in society is largely acquired through our families.
   4. Families provide ethnic identities and religious traditions which help form our cultural values and expectations.
F. There are several advantages to living in families: (1) families offer continuity as a result of emotional attachments, rights, and obligations; (2) families offer close proximity; (3) families offer familiarity; and (4) families provide many economic benefits by offering **economies of scale**.

V. EXTENDED FAMILIES AND KINSHIP
   A. The extended family consists not only of the cohabiting couple and their children but also of other relatives, especially in-laws, grandparents, aunts and uncles, and cousins.
   B. The social organization of the family, known as the **kinship system**, is based on the reciprocal rights and obligations of different family members.
      1. **Conjugal relationships** are extended family relationships created through marriage.
      2. **Consanguineous relationships** are created through biological (blood) ties.
      3. In some societies (mostly non-Western or nonindustrialized cultures), kinship obligations may be very extensive; close emotional ties between husband and wife are viewed as a threat to the extended family.
      4. The precedence of the kin group over the married couple is illustrated in the institution of **spirit marriage**.
      5. Despite the increasingly voluntary nature of kin relations, our kin create a rich social network for us.

VI. AN HISTORICAL PERSPECTIVE OF MARRIAGE AND THE FAMILY
   A. The greatest diversity in American family life probably existed during our country's earliest years.
      1. Over 240 groups of Native Americans, with distinct family and kinship patterns, inhabited what is now the U.S. and Canada.
      2. Many groups were **patrilineal**, others were **matrilineal**.
      3. Native American families tended to share certain characteristics.

B. European colonists who came to America attempted to replicate their familiar family system.
   1. This system (strongly influenced by Christianity) emphasized **patriarchy**, the subordination of women, sexual restraint, and family-centered reproduction.
   2. The family was basically an economic and social institution, the primary unit for producing most goods and caring for the needs of its members.
   3. Romantic love, not a factor in choosing a partner, came only after marriage and was considered a duty.
   4. The colonial family was strictly patriarchal.
      a. The authority of the husband/father rested in his control of land and property.
      b. A wife was not equal, but a "helpmate" who was economically dependent upon her husband.
      c. For women, marriage marked the beginning of a constant cycle of childbearing and childrearing.
      d. The colonial conception of childhood was radically different from today: Children were believed to be evil by nature.
C. During the eighteenth century and later, West African family systems were severely repressed throughout the New World.
   1. While unsuccessful at continuing polygamy, slaves were more successful at continuing traditions of strong extended family ties.
   2. Though legally prohibited, slaves created their own marriages.
   3. In the harsh slave system, the family provided strong support against the daily indignities of servitude.
D. In the nineteenth century, industrialization transformed American families from self-sufficient farm families to wage-earning urban families.
   1. A radically new division of family labor resulted.
   2. As "breadwinners", men's paid-in-wages work came to be identified as "real" work.
   3. As "housewives", women's unpaid work and services went unrecognized.
E. Without its central importance as a work unit, the family became the focus and abode of feelings.
   1. Love, as a basis for marriage, came to the foreground.
   2. Women now had a new degree of power in the ability to choose whom they would marry.
   3. The nineteenth century witnessed the most dramatic decline in fertility in American history: Women began to control the frequency of intercourse.
   4. A new sentimentality surrounded childhood and protecting children from the evils of the world became a major part of childrearing.
   5. In contrast to the colonial period, nineteenth-century adolescents were kept economically dependent and separate from adult activities.

F. Under slavery, the African-American family lacked two key factors which helped give free African-Americans and white families stability: autonomy and economic importance.
   1. The separation of slave families was common, creating grief and despair among thousands of slaves.
   2. It was impossible for the slave husband/father to become the provider for his family.
   3. When the formerly enslaved became free, the African-American family had strong emotional ties and traditions forged from slavery and their West African heritage.
G. In the nineteenth and early twentieth centuries, great waves of immigration swept over America.
   1. Most immigrants were uprooted: Leaving their homeland was never easy.
   2. Kinship groups were central to the immigrants' experience and survival.
   3. Many immigrant families could survive only by pooling their resources and sending mothers and children to work in the mills and factories.
H. By the beginning of the twentieth century, the functions of American middle-class families had been dramatically altered.
   1. Families lost many of their traditional economic, educational, and welfare functions.
   2. In the 1920's, a new ideal family form, based on **companionate marriage**, was beginning to emerge, rejecting the "old" family based on male authority and sexual repression.
   3. General and uncritical acceptance of traditional gender and marital roles prevailed during the golden age of the fifties.
   4. The fifties, as a time of unprecedented prosperity, fueled the movement to the suburbs which profoundly affected family life.

VII. READINGS AND FEATURES
   A. *Television and the World We Live In* is designed to help the reader get a sense of their "TV reality."
   B. In *Other Places ... Other Times: The Spirit Marriage in Chinese Society*, the authors describe the practice of spirit marriage, the arrangement of a marriage between two dead people.
      1. Spirit marriage guarantees that the male descent line will continue even in the face of death.
      2. The practice and meaning of spirit marriage in traditional Chinese society was shaped by cultural notions of kinship, marriage, and residence.
   C. *You and Your Well-Being: Defining Wellness* discusses the six components of well-being: physical, emotional, intellectual, spiritual, social, and environmental.
      1. All are interrelated and interconnected: A change in one influences the others.
      2. There is a strong connection between marriage and wellness.
   D. The *Perspective: Mapping Family Relationships: The Genogram* explains and illustrates the concept of a genogram.

Chapter 1

TEST YOUR COMPREHENSION

What is kin? Think of a person in your personal kinship network for each of the terms listed. Write each person's name under the column heading which describes their relationship to you. For example, if your best friend's name is Carol Mertens, write Carol Mertens under the "AFFILIATEDKIN" column. If no person fulfills the role, simply write the term in the column. When you finish, you will have an analysis of your own personal kinship network.

Chart 1.a

MY KINSHIP NETWORK		
FAMILY OF ORIENTATION	FAMILY OF COHABITATION	AFFILIATED KIN
sister Mom Dad	Lover Aunt cousin daughter	Betsy

aunt	foster child	lover	sibling
best friend	foster parent	member of the clergy	sister
boyfriend	girlfriend	mother	son
brother	godchild	mother-in-law	stepfather
cousin	godparent	neighbor	stepmother
daughter	grandfather	nephew	stepsibling
father	grandmother	niece	teacher
father-in-law	great-grandparent	second cousin	uncle
	half-siblings	pet	

List the four major family functions in the spaces provided below. Describe the traditional ways that these functions have been fulfilled and recent changes in these traditional functions.

Chart 1.b

FAMILY FUNCTION		
FUNCTION	TRADITIONAL FULFILLMENT	RECENT CHANGES
1. Christmas Family gathering — Distance		
2. Thanksgiving Family gathering — Distance		
3. Easter Family gathering — Distance		
4. New Year Family gathering — Distance		

The Meaning of Marriage and Family

Chapter 1

SELF-DISCOVERY I

Examine your personal kinship network. Select the most important female role model. What is her relationship to you? How does she affect your attitude regarding women and women's roles? How does she affect your attitude regarding men and men's roles?

Select the most important male role model. What is his relationship to you? How does he affect your attitude regarding men and men's roles? How does he affect your attitude regarding women and women's roles?

The Meaning of Marriage and Family

SELF QUIZZES

How well do you know this material? Test your understanding of Chapter One by answering the following sample questions.

Part I - Multiple Choice: Choose the most correct response.

__d__ 1. Which of the following statements is **not** true regarding American television?
   a. It is probably the most pervasive medium in our culture.
   b. It provides models which guide our interactions with others.
   c. Heavy television viewers tend to have different beliefs and attitudes about the world than light television viewers.
   d. Most television programming presents a realistic view of American families.
   e. Television makes us privy to behaviors we may not see publicly.

__e__ 2. Cultivation theory is related to
   a. consistent stereotypes.
   b. a consistent world view.
   c. media research.
   d. images of men and women.
   e. all of the above

__a__ 3. Television sitcoms
   a. portray working-class and middle-class families differently.
   b. avoid marital stereotypes.
   c. represent ethnic families on an equally proportionate basis as white families.
   d. reflect the complexity of living in real marriages and families.
   e. portray the reality that most sexual interactions take place within marriage.

__a__ 4. Single-parent families and stepfamilies on television
   a. are usually formed as a result of a spouse's death.
   b. are usually formed as a result of a divorce.
   c. are usually formed as a result of a birth to an unmarried woman.
   d. present a realistic viewpoint of family dynamics compared to intact families.
   e. both a and d

__b__ 5. The preferred marital arrangement worldwide is
   a. monogamy.
   b. polygamy.
   c. serial monogamy.
   d. polyandry.
   e. modified polygamy.

11

Chapter 1

_e_ 6. Which of the following might be considered affiliated kin?
   a. Latinos' compadres
   b. your second cousin
   c. Japanese Americans' ie
   d. your best friend
   e. both a and d

_e_ 7. The need for intimacy
   a. is a primary human need.
   b. may keep unhappy marriages together.
   c. strongly influences health and mortality rates.
   d. may be fulfilled by pets.
   e. all of the above

_c_ 8. Which of the following functions has shifted dramatically away from the family?
   a. intimacy
   b. economic cooperation
   c. socialization
   d. archival
   e. status assignment

_e_ 9. One advantage of living in families is related to
   a. proximity.
   b. economies of scale.
   c. continuity.
   d. familiarity.
   e. all of the above

_c_ 10. Consanguineous relationships
   a. are created through marriage.
   b. are created through spirit marriage.
   c. are created through biological (blood) ties.
   d. are affiliated kin relationships.
   e. both b and c

_e_ 11. Colonial families
   a. were strongly rooted in patriarchy.
   b. considered the importance of love in marital choice.
   c. were the primary unit for producing most goods and caring for the needs of its members.
   d. believed children were evil by nature.
   e. all but b

The Meaning of Marriage and Family

__b__ 12. As a result of industrialization,
   a. American families became more self-sufficient.
   (b.) men's work was given higher status than women's work.
   c. the family became more important as a work unit.
   d. women lost power in terms of marital choice and frequency of intercourse.
   e. children and adolescents participated in the adult world of work and other activities.

__e__ 13. African-American slave families
   a. lacked autonomy and economic importance.
   b. were commonly separated.
   c. valued extended kin.
   d. had strong emotional ties and traditions.
   (e.) all of the above

__e__ 14. Companionate marriages and families
   a. began to emerge shortly after the Civil War.
   b. were based on male authority and sexual repression.
   c. protected children from the world.
   d. emphasized patriarchal decision-making.
   (e.) expected marriages to provide romance, sexual fulfillment, and emotional growth.

Part II - True/False

__T__ 1. Popular culture is one of our key sources of misinformation.

__F__ 2. Themes which run through sitcoms usually present an accurate picture of the realities of married life and/or family living.

__T__ 3. Relatively little marital sex is depicted on television although most sexual interactions take place within the context of marriage.

__F__ 4. Latinos and Asian Americans are adequately represented on American television.

__F__ 5. Eighty-four percent of the world's cultures prefer monogamy as a marriage style.

__T__ 6. With the current high divorce and remarriage rate, our marriage system might best be called "serial monogamy" or "modified polygamy."

__F__ 7. In the affiliated family, a person is considered a family member only if he/she is related by blood.

Chapter 1

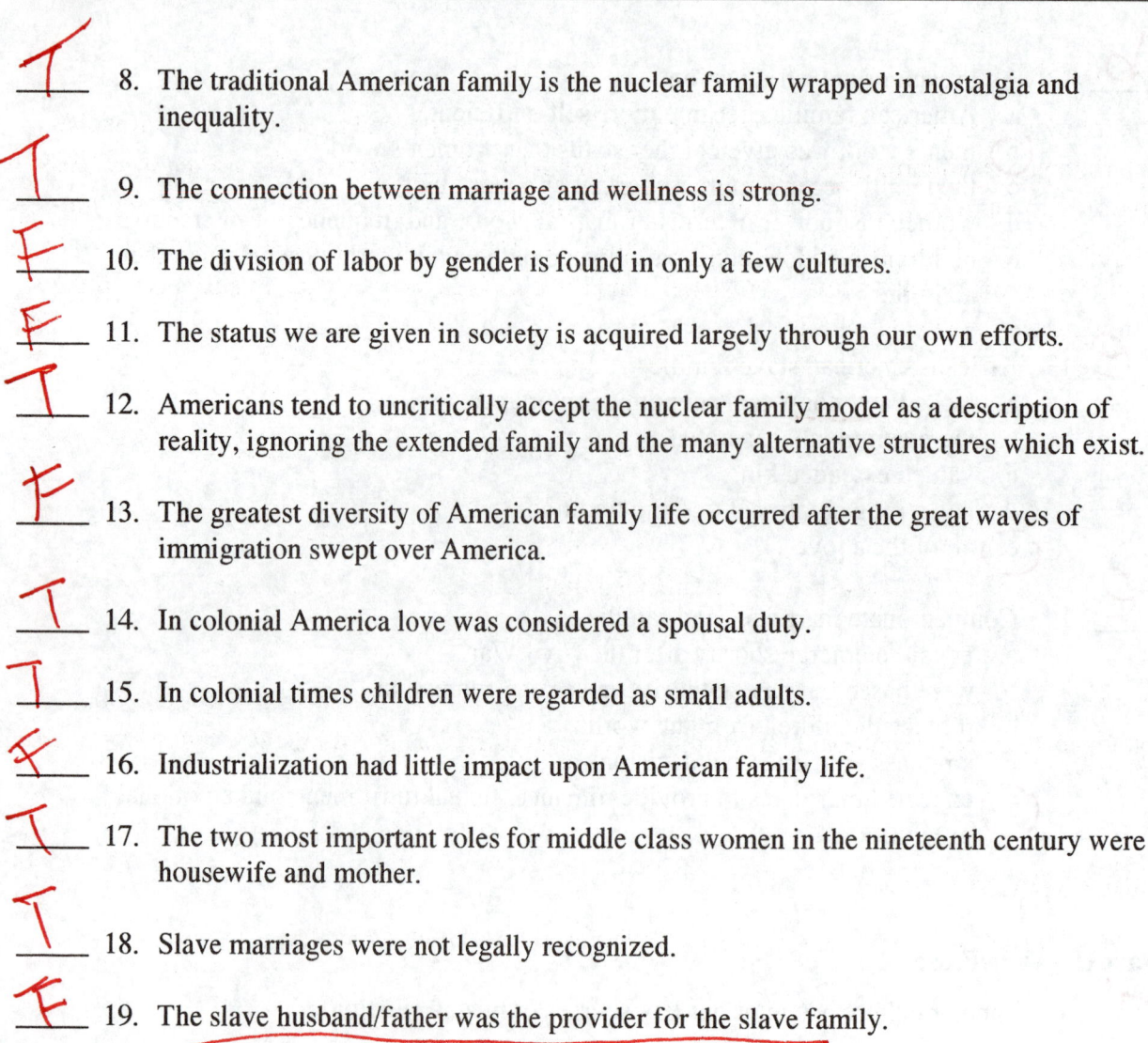

__T__ 8. The traditional American family is the nuclear family wrapped in nostalgia and inequality.

__T__ 9. The connection between marriage and wellness is strong.

__F__ 10. The division of labor by gender is found in only a few cultures.

__F__ 11. The status we are given in society is acquired largely through our own efforts.

__T__ 12. Americans tend to uncritically accept the nuclear family model as a description of reality, ignoring the extended family and the many alternative structures which exist.

__F__ 13. The greatest diversity of American family life occurred after the great waves of immigration swept over America.

__T__ 14. In colonial America love was considered a spousal duty.

__T__ 15. In colonial times children were regarded as small adults.

__F__ 16. Industrialization had little impact upon American family life.

__T__ 17. The two most important roles for middle class women in the nineteenth century were housewife and mother.

__T__ 18. Slave marriages were not legally recognized.

__F__ 19. The slave husband/father was the provider for the slave family.

__T__ 20. Most immigrant groups experienced hostility in America.

__T__ 21. General and uncritical acceptance of traditional gender and marital roles prevailed during the 1950s.

The Meaning of Marriage and Family

Part III - Matching

The following terms can easily be remembered if you carefully examine the root word or words. The root word for marriage is "gamy" and is combined with other forms. "Mono" equals "one" (monocle, monarchy). "Bi" is the root for "two" (bicycle, bisexual). "Poly" is "many" or "plural" (polymer). "Gyn" is a root word for "woman" or "female" (gynecologist, misogyny), while "andro" is a root word for "man" or "male" (android). Using this etymology, match each of the following terms to its definition:

_b_ 1. polygamy   a. having only one spouse at a time
_d_ 2. polygyny   b. having more than one spouse at a time
_e_ 3. polyandry   c. having one spouse after another
_a_ 4. monogamy   d. having more than one wife at a time
_c_ 5. serial monogamy   e. having more than one husband at a time
_f_ 6. bigamy   f. having two husbands or wives (illegally)

Food for Thought: If the plural of mouse is mice, is the plural of spouse therefore spice? Is monogamy equal to monotony? If marriage is a great institution, do you really want to live in an institution?

DISCUSS BRIEFLY

1. How have the changing family functions led to an increase in the divorce rate?

2. Many people assume that the high divorce rate indicates poor family relationships and increased personal unhappiness. Would the authors of your text agree with this assessment? Why or why not?

Chapter 1

MINI-ASSIGNMENT I

Ask a grandparent or person over 60 years old their opinion regarding what the roles of men and women should be. What does he/she think are the major problems facing the modern family and what solutions could be offered?

Person #1_____   Age_____

Opinion:

Ask one or both of your parents (or another significant person from your family of orientation) the same questions.

Person #2_____   Age_____

Opinion:

Do you agree or disagree with these assessments? Why or why not?

MINI-ASSIGNMENT II

On a large piece of paper (space provided here would be insufficient), map out your own genogram, as discussed in the text. Using the text as a model, draw a genogram including information about you, your parents and your grandparents. Be sure to include:

- Name and date of birth. If deceased, year and cause of death.
- Education and occupation.
- Siblings in each generation.
- Marital history, including years of marriage, divorce, dates of remarriages, if any.
- Health or psychological problems.
- The nature and quality of emotional relationships between family members.
- Assigned role each member played in the family.

SELF-DISCOVERY II

Your text stresses the importance of intimacy as a function of the family. This supports Abraham Maslow's idea that the need for belonging is secondary only to our safety and security needs. Why do you think so many individuals in our society can lead such hectic lives and still remain lonely? Is loneliness always bad or can it be positive as well? Why are so many people lonely and yet hesitate to turn to each other?

Chapter 1

SELF-DISCOVERY III

Why have you chosen to take a course such as the one using this text? Write down the personal goals which you hope to address through this class.

1.

2.

3.

What are you willing to do to address these goals?

1.

2.

3.

KEY TO SELF QUIZZES

Multiple Choice

1. d
2. e
3. a
4. a
5. b
6. e
7. e
8. c
9. e
10. c
11. e
12. b
13. e
14. e

True/False

1. T
2. F
3. T
4. F
5. F
6. T
7. F
8. T
9. T
10. F
11. F
12. T
13. F
14. T
15. T
16. F
17. T
18. T
19. F
20. T
21. T

Matching

1. b
2. d
3. e
4. a
5. c
6. f

SUGGESTED READINGS

For related readings, turn to pages 36 - 37 in the text.

CHAPTER 2

Studying Marriage and Family

MAIN FOCUS

Chapter Two examines thinking critically about marriage and the family, research methods, theories of marriage and the family, ethnicity and family research, and contemporary American marriages and families. It offers insights into approaching and understanding the family from a scientific perspective.

GOALS OF THIS CHAPTER

To demonstrate mastery of this chapter, you should be able to:

1. Explain the importance of studying marriage and the family.
2. Define objectivity and discuss ways of thinking that lack objectivity.
3. Understand and explain the prevalence and impact of the advice/information genre.
4. Describe social science research methods and methodology.
5. Recognize the strengths and weaknesses of survey research, clinical research, observational research and experimental research.
6. Understand and explain the major assumptions, viewpoints and criticisms of symbolic interactionism, social exchange theory, structural functionalism, conflict theory, family systems theory, and family development theory.
7. Discuss the orienting focus of the feminist perspective.
8. Differentiate between ethnic group, racial group, and minority group.
9. Explain the impact of research and/or lack of research upon our views of ethnic families.
10. Recognize important features of African-American, Latino, Asian-American, and Native-American families.

Chapter 2

11. Understand and explain white ethnicity.
12. Explain dramatic changes which have occurred related to contemporary American marriages and families.
13. Understand the Circumplex Model of Family Functioning.

KEY TERMS AND IDEAS

The following terms, ideas, and concepts are listed in the order in which they appear in Chapter Two. Be sure that you understand and can define each of the following:

objectivity	case study method	family systems theory
values	observational research	homeostasis
value judgments	experimental research	feminist perspective
objective statements	variables	gender
opinions	independent variables	family development theory
biases	dependent variables	ethnic group
stereotypes	correlational studies	racial group
fallacies	theory	phenotype
egocentric fallacy	symbolic interaction	minority group
ethnocentric fallacy	interaction	status
norms	symbols	socioeconomic status
advice/information genre	social role	familialism
scientific method	social exchange theory	extended households
quantitative research	structural functionalism	cohabitation
qualitative research	subsystems	separation
secondary data analysis	instrumental trait	divorce
survey research	expressive trait	extended family
clinical research	conflict theory	

CHAPTER TWO OUTLINE

I. INTRODUCTION TO THE CHAPTER
   A. While it is good to know about your own family experiences, effective planning and decision-making requires that we also have a larger background of information.
   B. The study of family is valuable because it promotes a more informed understanding which helps people in terms of their individual lives and helps social service, medical, and legal personnel as they deal with family-related issues.

II. THINKING CRITICALLY ABOUT MARRIAGE AND THE FAMILY
   A. In order to obtain valid research information, researchers and research consumers need to keep in mind the rules of critical (clear and unbiased) thinking.
   B. Personal experience creates personal perspectives, values, and beliefs which can create blinders that keep people from accurately reading research information.
   C. **Objectivity** in approaching information means that we suspend the beliefs, biases, or prejudices we have about a subject until we really understand what is being said, then relating it to the information and attitudes we already have.
   D. The **values** we have about what makes a successful family can cause us to decide ahead of time that certain family life styles are abnormal.
   E. **Opinions**, **biases**, and **stereotypes** are ways of thinking that lack objectivity.
   F. **Fallacies** are errors in reasoning.
      1. **Egocentric fallacies** are mistaken beliefs that everyone has the same experiences and values that we have and therefore should think as we do.
      2. **Ethnocentric fallacies** are beliefs that one's own ethnic group, nation, or culture is innately superior to others.

III. CONTEMPORARY AMERICAN MARRIAGE AND FAMILIES
   A. Apart from our own families, popular culture may be the most important vehicle through which **norms** (cultural rules and standards) and **values** about marriage and the family are transmitted.
   B. The **advice/information genre** transmits norms and information through books, advice columns, radio and television shows, and newspaper and magazine articles.
   C. The advice/information genre, apparently concerned with transmitting information that is factual and accurate, produces columns, articles, and programs which share the following features:
      1. Their primary purpose is to sell a product (e.g. newspapers) or to raise program ratings.
      2. The media must entertain while disseminating information about marriage, family, and relationships.
      3. The material focuses on how-to-do-it information or morality.
      4. The genre uses the trappings of social science without its substance.
   D. Guidelines for evaluating the advice/information genre include: be skeptical; look for biases, stereotypes, and lack of objectivity; look for moralizing; go to the original source(s); and seek additional information.

IV. RESEARCH METHODS
   A. Family research is the process of bringing together information and formulating generalizations about certain areas of experience.
      1. Our individual research process is vital to our effective coping with life and to our understanding of life's meaning, however, it may limit our interest in obtaining further background and insights.

Chapter 2

        a. The scientific information in this text provides an opportunity to consider one's present attitudes and past experiences in relation to others and one's future.
        b. Effectively applying what is relevant to one's life requires the use of logic and the problem solving skills of critical thinking.
  B. Family science researchers use **scientific method**, well-established procedures to collect information about family experiences.
    1. **Quantitative research** involves asking the same questions to a great number of persons using representative sampling.
    2. **Qualitative research** involves studying smaller groups or individuals in a more in-depth fashion using intensive interviews, case studies, or various documents.
    3. **Secondary data analysis** involves reanalyzing data originally collected for another purpose.
  C. Family researchers conduct their investigations using ethical guidelines agreed upon by professional researchers: These guidelines protect the privacy and safety of research participants and assure the trustworthiness of their research reports.
  D. **Survey research** (using questionnaires or interviews) is the most popular data-gathering technique in marriage and family studies.
    1. Survey research is designed to gather information from a small, representative group and to infer conclusions that are valid for a larger population.
    2. Questionnaires offer anonymity, may be quick to complete, and are relatively inexpensive; however, they do not allow for an in-depth response.
    3. Interviews have more depth, but are open to subjective interpretation.
    4. Surveys have inherent problems:
        a. They must have a representative sample.
        b. The subjects may not really understand their own behavior.
        c. People tend to under-report undesirable behavior.
  E. **Clinical research** involves in-depth examinations of persons or small groups who go to psychiatrists, psychologists or social workers with problems.
    1. The **case-study method** is the most traditional approach to clinical research.
    2. Clinical studies offer long-term, in-depth study of various aspects of marriage and family; however, such individuals may not represent the general population.
    3. Clinical research has been fruitful in developing insights into family processes.
  F. In **observational research**, scholars attempt to study behavior systematically through direct observation while remaining as unobtrusive as possible.
    1. An obvious disadvantage of using this method is that people may hide unacceptable ways of dealing with decisions while an observer is present.
    2. Self-reports differ from observations and may measure two different views of the same thing.
  G. In **experimental research**, researchers isolate a single factor under controlled circumstances to determine its influence.

1. Researchers control experiments by using **variables**.
2. Two types of **variables** are **independent variables** (factors that can be manipulated or changed by researcher) and **dependent variables** (factors that are affected by changes in the independent variable).
3. **Correlational studies** (clinical studies, surveys, and observational research) measure two or more naturally occurring variables to determine their relationship to each other.
4. Experimental findings can be very powerful, because such research gives investigators control over many factors and enables them to isolate variables.
5. A problem with experimental research is that people respond differently to others in real life than in a controlled situation.

V. THEORIES OF MARRIAGE AND THE FAMILY
  A. A **theory** is a set of general principles or concepts used to explain data and to make predictions that may be empirically tested.
  B. **Symbolic interactionism** looks at how people interact with one another, communicating with symbols and gestures.
    1. **Interaction** is a reciprocal act that takes place between people and uses **symbols**.
    2. The family can be seen as a unity of interacting personalities, with each member having a **social role**.
      a. Over time, our interactions and relationships define the nature of our family.
      b. Our identities emerge from the interplay between our unique selves and our social roles.
    3. Symbolic interactionism has several weaknesses.
      a. Symbolic interaction tends to minimize the role of power in relationships.
      b. It does not account for the psychological aspects of life, emphasizes individualism, and does not place marriage or family within a larger social context.
  C. **Social exchange theory** examines actions and relationships in terms of costs and benefits.
    1. Much of this cost-benefit analysis is unconscious.
    2. In personal relationships, resources, rewards, and costs are more likely to be things like love, companionship, status, power, fear and loneliness, rather than tangibles, such as money.
    3. People consciously or unconsciously use their various resources to obtain what they want.
    4. Exchanges which occur have to be fair and must have equity: Both partners feel uneasy in an inequitable relationship.
    5. Because marriages are expected to endure, exchanges take on a long-term character and are either cooperative or competitive.

Chapter 2

      6. Problems with social exchange theory include: it assumes rationality when humans are not always rational; it has difficulty ascertaining the value of costs and rewards; and values which are assigned are highly individualistic.
- D. **Structural functionalism** theorizes about how society works, how families work, and how families relate to larger society and to their own members.
  1. The family is viewed as the **subsystem** of society which provides new members for society through procreation and socialization.
  2. It examines how the family organizes itself for survival and what functions the family performs for its members.
  3. Structural functionalists encourage men to develop **instrumental traits** and women to develop **expressive traits**.
  4. Criticisms of this theory include: (1) it cannot be tested empirically; (2) it is not always clear what function a particular structure serves; (3) it has a conservative bias against change; and (4) it looks at the family abstractly and has little relevance to real families in the real world.
- E. **Conflict theory** maintains that life involves discord; society is divided rather than cooperative.
  1. In addition to love and affection, conflict theorists believe that conflict and power are fundamental to marriage and family relationships.
  2. Conflict theorists view conflict as a natural part of family life and not necessarily bad.
  3. Conflict theory recognizes four sources of power: legitimacy, money, physical coercion, and love.
  4. Everyone in the family has power although the sources and degree of power may vary.
  5. Conflict theory seeks to channel conflict and to search for solutions through communication, bargaining, and negotiations.
  6. Criticisms of conflict theory include: (1) it fails to recognize the power of love or bonding; (2) it assumes differences lead to conflict; and (3) conflict in families is not easily measured or evaluated.
- F. **Family systems theory** sees the family as a structure of related subsystems: Each subsystem carries out certain functions.
  1. An important task of subsystems is maintaining boundaries: When the boundaries become blurred, the family becomes dysfunctional.
  2. Interactions are important in family systems theory.
  3. Family systems therapists and researchers believe: (1) interactions must be studied in the context of the family system; (2) the family has a structure only visible in its interactions; (3) the family is a purposeful system which seeks **homeostasis**; and (4) despite resistance to change, each family system is transformed over time.
  4. Many of the basic concepts of family systems theory are still in dispute.

G. The **feminist perspective** is not a unified theory. Gender differences are the orienting focus in most feminist writing, research, and advocacy.
1. Feminists maintain family and gender roles have been socially constructed as ways by which men maintain power over women.
2. Feminists urge a more extended view of family to include all kinds of sexually interdependent adult relationships regardless of the legal, residential, or parental status.
3. Feminists campaign to raise society's level of awareness to the oppression of women and associate their concern for greater sensitivity to all disadvantaged groups.
4. The feminist agenda is to attend to the social context as it impacts personal experience and to work to translate personal experience into community action and social critique.
5. The feminist perspective includes a variety of viewpoints. These viewpoints have an integrating focus relating to power and the inequity of power prevailing in the positions of men and women in society and especially in family life.

H. **Family development theory** examines the changes in the family beginning with marriage and proceeding through seven more stages.
1. The lives of all people involve response to certain universal developmental challenges.
2. The life-cycle model gives insight into the complexities of family life, the different tasks families perform, and changing roles and circumstances over time.
3. Family development theory provides a framework with which to view the maturational development of individuals as it influences and is influenced by the social environment of the family.

VI. ETHNICITY AND FAMILY RESEARCH
A. An **ethnic group** is a group of people distinct from other groups due to cultural characteristics (e.g. language, religion, customs) that are transmitted from one generation to another.
B. A **racial group** is a group of people classified according to their phenotype which is determined by anatomical and physical characteristics, such as skin color and facial structure.
C. A **minority group** is a group of people whose **status** places them at an economic, social, or political disadvantage.
D. Until the last twenty years, most research about American marriages and families was limited to the white, middle-class nuclear family.
1. Instead of recognizing the strengths of African-American, Latino, Asian-American, and Native-American families, we have viewed them as "tangles of pathology" for failing to meet the model of the traditional nuclear family.
2. Two of the most prominent examples of ethnocentric distortions are the "culture of poverty" approach to studying African-American families and the "machismo syndrome" approach to studying Latino families.

Chapter 2

 E. The largest ethnic group in the United States is African-Americans.
   1. African-American families have a strong sense of **familialism** which emphasizes family loyalty and kinship.
   2. Understanding **socioeconomic status** (rank in society based on a combination of occupational, educational, and income levels), especially poverty, is critical to studying African-American life.
   3. Striking features of African-American families include: a long history of dual-earner families as a result of economic need (creating more egalitarian family roles than white families); the importance of kinship bonds; a strong tradition of familialism; the fact that children are highly valued; and the likelihood of living in **extended households.**
 F. Latinos are the fastest growing and second largest ethnic group in America.
   1. There is considerable diversity among Latinos in terms of ethnic heritage and socioeconomic status.
   2. Important factors in Latino family life include close kin cooperation and mutual assistance, the importance of children, Catholicism, and a tradition of male dominance (although day-to-day living patterns suggest that women have considerable power and influence in the family).
 G. Asian-Americans are a particularly diverse group.
   1. The largest Asian-American groups are Chinese-Americans, followed by Filipino-Americans and then Japanese-Americans.
   2. More recent arrivals (Vietnamese, Cambodians, Laotians, and Hmong) first immigrated to the United States in the 1970s as refugees of the Vietnam War.
   3. Values that continue to be important to Asian-Americans include the importance of family over the individual, self-control to achieve societal goals, and appreciation of one's cultural heritage
   4. The most dramatic change affecting Chinese-Americans has been their striking increase in population over the past twenty years.
    a. Because of the large numbers of new immigrants, it is important to distinguish between American-born and foreign-born Chinese-Americans.
    b. Contemporary American-born Chinese families: emphasize familialism; tend to be better educated, have higher incomes and lower rates of unemployment than the general population; have conservative sexual values and attitudes toward gender roles; have a strong sense of family; and expect women to be employed and contribute to household income.
 H. Approximately 2 million Americans identify themselves as being of native descent: Those deeply involved in their own traditional culture give themselves a tribal identity while those who are more acculturated tend to give themselves an ethnic identity.
   1. There has been a considerable migration of Native-Americans to urban areas since World War II because of poverty on reservations and pressures toward acculturation.
   2. Because of the importance of tribal identities and practices, there is no single type of Native-American family.

3. Although considerable variation exists among different tribal groups, extended families are significant to Native-Americans and large numbers of Native-Americans are married to non-Native-Americans.

I. In recent years, the sense of ethnicity has been growing among Americans of European descent, especially among the working-class.
   1. White ethnicity is strongest in the East and Midwest.
   2. Symbolic ethnicity is an ethnic identity used only when the individual chooses.
   3. Researchers have concluded that a European/non-European distinction remains a central division in our society because most European ethnic groups no longer have **minority status** and most European ethnic groups are not physically distinguishable from other white Americans.

VII. CONTEMPORARY AMERICAN MARRIAGES AND FAMILIES
   A. **Cohabitation** refers to relationships in which unmarried individuals share living quarters and are sexually involved.
   B. A number of factors have altered the meaning of marriage and the role it plays.
   C. **Separation** occurs when two married people no longer live together.
   D. **Divorce** is the legal dissolution of marriage.
      1. Divorce has become so widespread that many scholars are beginning to consider it one variation of the normal life course.
      2. Contemporary divorce patterns create three common experiences for American marriages and families: single-parent families, remarriages, and stepfamilies.
      3. **Extended families** include in-laws, grandparents, aunts and uncles, and cousins.

VIII. Readings and Features
   A. *Understanding Yourself* examines what surveys tell you about yourself: Survey questionnaires are the leading source of information about family and marriage.
   B. The **Family Circumplex Model** is an important tool for identifying and mapping family relationships.
      1. Three important components of family dynamics are **cohesion, adaptability**, and **communication**.
      2. **Enmeshment** occurs when individuals overidentify with the family.
      3. **Disengagement** occurs when family members do not feel close and do not communicate.
      4. Balanced families (those with moderate levels of cohesion and adaptability) are seen as having the greatest marital and family strengths across the family life cycle.
   C. *Other Places... Other Times*: In many ways, the Ohlone in California reflected many of the basic characteristics of Native-American groups.

Chapter 2

TEST YOUR COMPREHENSION

Chart 2.A compares the primary research methods used in studying the family. Complete the chart, describing the method, advantages, and limitations of each type of research.

Chart 2.A

PRIMARY RESEARCH METHODS	
METHOD	DESCRIPTION OF METHOD
Questionnaire Surveys — data-gathering technique in marriage and family studies	gather information from a small group and infer conclusions that are valid for a larger population.
Interview Surveys	any / Both
Clinical Research — indepth examinations of persons or small group who go to psychiatrists, psychologist or social workers w/ problems.	
Observational Research — Scholars attempt to study behavior systematically through direct observation while remaining as unobtrusive as possible.	
Experimental Research — Researchers isolate a single factor under controlled circumstances to determine its influence.	1. Researchers control experiments by using variables. 2. Two types of variables are independent variables

28

IN MARRIAGE AND FAMILY

ADVANTAGES OF METHOD	LIMITATIONS OF METHOD
Questionnaires offer anonymity, may be quick to complete & are inexpensive; they do not allow for an in-depth response. Interviews have more depth, but are open to subjective interpretation.	They must have a representative sample. The subject may not really understand their own behavior. People tend to under-report undesireable behavior.
The case-study method is the most traditional approach to clinical research. Clinical research has been fruitful in developing insights into family processes.	Clinical studies offer long-term in-depth study of various aspects of marriage and family; however, such individuals may not represent the general population.
	An obvious disadvantage of using this method is that people may hide unacceptable ways of dealing with decisions while an observer is present. observational research
Self-reports differ from observations and may measure two different views of the same thing.	
Correlation studies, clinical studies, surveys, and observation research. Experimental findings can be very powerful, because such research gives investigators control over many factors and enables them to isolate variables.	A problem with experimental research is that people respond differently to others in real life than in a controlled situation.

Chapter 2

Chart 2.B is a chart contrasting the seven theories discussed in your text: Symbolic Interaction, Social Exchange Theory, Structural Functionalism, Conflict Theory, Family Systems Theory, Feminist Perspective, and Family Development Theory. Complete the charts, using your text, lecture notes and outside reference material if necessary.

Chart 2.B

THEORIES OF MARRIAGE		
THEORY	DESCRIPTION	ASSUMPTIONS (KEY BELIEFS)
Symbolic Interaction — look at how people interact with one another, communication with symbols and gestures.	The family interacting having a social role.	can be seen as unity of personalities with each other.
Social Exchange Theory — examines actions and relationships in terms of cost and benefits which analysis is unconscious.	In personal relationships, resources/rewards are more likely to be things like love, companionship, status, power, fear, + loneliness.	
Structural Functionalism — theories about society and relationships	works, how families work and how families	relate to larger society and to their own members. encourage women to develop instrumental traits and men to develop expressive traits.
Conflict Theory — maintains that life involves discord, society is divided rather than cooperative	The believe that conflict and power are fundamental to marriage and family relationships.	Natural conflict is not bad but a natural part of family life.
Family Systems Theory — sees the family as a structure of related subsystems	Maintaining boundaries & when they become blurred, the family can become dysfunctional	Interactions between family. Family systems therapists / researchers believe
Feminist Perspective — is not unified theory, gender differences are the orienting focus	In most feminist writing, research, advocacy	
Family Development Theory — examines the changes in the family in marriage, & stages of		

30

AND THE FAMILY

UNIQUE TERMINOLOGY	VIEWS ON GENDER	LIMITATIONS weaknesses
Over time our interactions and relationships define the nature of our family.	Our identities emerge from the interplay between our unique selves and our social roles.	① Symbolic interaction tends to minimize the role of power in relationships. ② It does not account for the psychological aspects of life, emphasizes individualism, and does not place marriage or family within a larger social context.
~~Exchange~~ fair exchange ⟨Problems - humans are not always rational⟩	Because marriages are expected to endure a long term, character exchanges are either cooperative or competitive. see pg 24	
The family is viewed as ~~to take care of the~~ ~~our members~~ society which provides new members for society through procreation & socialization	the subsystem of pg 24	It cannot be tested empirically. 2.
conflict theory recognizes power ① legitimacy, money, physical ② channel conflict to research for solutions through communication, bargaining, & negotiation.	four sources of power coercion, love,	① It fails to recognize the power of love or bonding. ② it assumes that difference lead to conflict; ③ conflict in family is not easily measured or evaluated.
① Study Interactions ② family structure only visible in it interactions.	see pg 24	
① Maintain power over women.		
life cycle - complex maturational development as it influences the family.	development in individuals social environment of	

Chapter 2

SELF QUIZZES

How well do you know this material? Test your understanding of Chapter Two by answering the following sample questions.

Part I - Multiple Choice: Choose the most correct response.

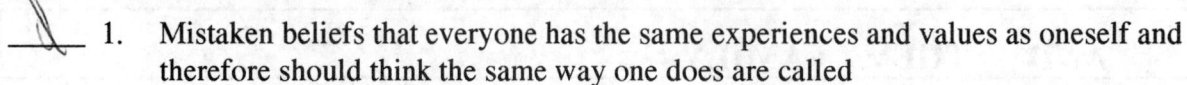

___ 1. Mistaken beliefs that everyone has the same experiences and values as oneself and therefore should think the same way one does are called
   a. ethnocentric fallacies.
   b. egotistical fallacies.
   c. ethnologic fallacies.
   d. egocentric fallacies.
   e. none of the above

___ 2. Studying small groups or individuals in an in-depth fashion using intensive interviews or case studies is a scientific method known as
   a. secondary analysis.
   b. qualitative research.
   c. comprehensive analysis.
   d. quantitative research.
   e. empirical documentation.

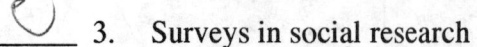

___ 3. Surveys in social research
   a. represent only a small source of our information.
   b. usually allow for an in-depth response to issues.
   c. gather information from a small representative group of people and infer conclusions valid for a larger population.
   d. have recently decreased in usage.
   e. are a good indicator of how well people interact with one another.

___ 4. Which of the following is **not** a major problem in using the survey method in family research?
   a. In interviews, researchers may allow their own preconceptions to bias their interpretations.
   b. The sample that volunteered to take the survey may not be representative.
   c. Research participants may not understand their own behavior.
   d. Questionnaires are one of the most expensive and difficult research methods.
   e. Research participants under-report undesirable behavior.

___ 5. Clinical studies have been very beneficial in
   a. providing representative samples.
   b. making inferences to the population in general.
   c. developing insights into family processes.
   d. developing family systems theory.
   e. both c and d

Studying Marriage and Family

__a__ 6. Which of the following types of research is **not** correlational in nature?
   a. experimental research
   b. clinical studies
   c. observational research
   d. survey research
   e. All of the above are correlational.

__a__ 7. Symbolic interaction theory
   a. examines how people interact, communicate, and are socialized.
   b. looks at how each individual functions within the family unit.
   c. examines resources and power.
   d. focuses on the exchange between people who love each other.
   e. examines the unconscious processes of emotions.

__a__ 8. Social exchange theory focuses on
   a. measuring our relationships on a cost/benefit basis.
   b. the pleasure of fulfilling our ascribed role.
   c. wealth and power used to control others.
   d. harmony and stability maintained in relationships.
   e. problem solving and resolution of conflict.

__d__ 9. The theory which explains how society works is
   a. social exchange theory.
   b. family systems theory.
   c. conflict theory.
   d. structural functionalism.
   e. symbolic interactionism.

__d__ 10. Which of the following is **not** characteristic of conflict theory as applied to the family?
   a. It sees the family in terms of conflicts of interest.
   b. Conflict within the family is seen as natural and not necessarily bad.
   c. Family sources of power include legitimacy, money, physical coercion, and love.
   d. It focuses on love, bonding, and cooperation.
   e. Everyone in the family has power, although the power may be different and unequal.

__e__ 11. Family systems theorists
   a. focus on the patterns of interaction between various family members.
   b. view the family as a structure of related parts or subsystems.
   c. see subsystems with blurred boundaries as dysfunctional.
   d. believe family systems transform over time
   e. all of the above

Chapter 2

_____ 12. The feminist perspective maintains that traditional gender roles
   a. are a social construction.
   b. create oppressive conditions and barriers to opportunity.
   c. result from biological conditions.
   d. have been created to maintain the power of men over women.
   e. all but c

_____ 13. A group of people who share common phenotypical characteristics is known as a
   a. minority group.
   b. racial group.
   c. phenotypical group.
   d. ethnic group.
   e. collective group.

_____ 14. Most research about American marriages and families has been conducted on
   a. African-Americans.
   b. Japanese-Americans.
   c. White, middle class Americans.
   d. Asian-Americans.
   e. Native-Americans.

_____ 15. The largest ethnic group in the United States is
   a. Asian-Americans.
   b. Latinos.
   c. African-Americans.
   d. Native-Americans.
   e. none of the above

_____ 16. Better education and higher incomes than the general population tend to be found in which of the following ethnic groups?
   a. African-Americans
   b. Chinese-Americans
   c. Native-Americans
   d. Latinos
   e. Japanese-Americans

_____ 17. Which of the following no longer has a minority status?
   a. Native-Americans
   b. African-Americans
   c. Latinos
   d. European ethnic groups
   e. Asian-Americans

_____ 18. The primary purpose of the advice/information genre is
   a. pursuing knowledge.
   b. selling newspapers or magazines.
   c. raising program ratings.
   d. different than scholarly research.
   e. all but a

Studying Marriage and Family

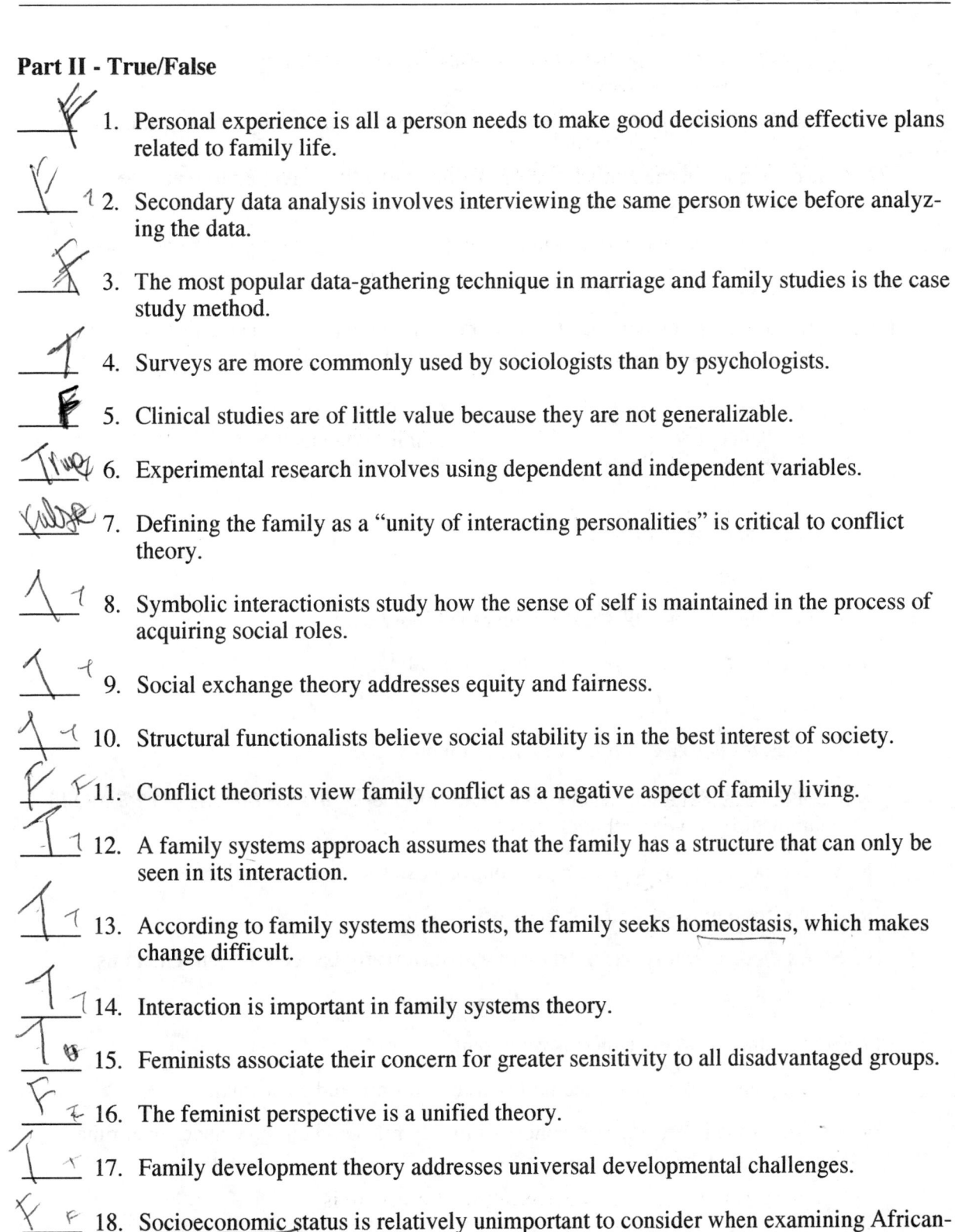

Part II - True/False

___F___ 1. Personal experience is all a person needs to make good decisions and effective plans related to family life.

___F___ 2. Secondary data analysis involves interviewing the same person twice before analyzing the data.

___F___ 3. The most popular data-gathering technique in marriage and family studies is the case study method.

___T___ 4. Surveys are more commonly used by sociologists than by psychologists.

___F___ 5. Clinical studies are of little value because they are not generalizable.

___True___ 6. Experimental research involves using dependent and independent variables.

___False___ 7. Defining the family as a "unity of interacting personalities" is critical to conflict theory.

___T___ 8. Symbolic interactionists study how the sense of self is maintained in the process of acquiring social roles.

___T___ 9. Social exchange theory addresses equity and fairness.

___T___ 10. Structural functionalists believe social stability is in the best interest of society.

___F___ 11. Conflict theorists view family conflict as a negative aspect of family living.

___T___ 12. A family systems approach assumes that the family has a structure that can only be seen in its interaction.

___T___ 13. According to family systems theorists, the family seeks homeostasis, which makes change difficult.

___T___ 14. Interaction is important in family systems theory.

___T___ 15. Feminists associate their concern for greater sensitivity to all disadvantaged groups.

___F___ 16. The feminist perspective is a unified theory.

___T___ 17. Family development theory addresses universal developmental challenges.

___F___ 18. Socioeconomic status is relatively unimportant to consider when examining African-American families.

___F___ 19. The fastest growing ethnic group in the United States is Asian Americans.

Chapter 2

___ 20. Latino women have little power and influence in their families.

___ 21. Asian-Americans value a strong sense of importance of family over the individual.

___ 22. The endurance of the Native-American ethnic identity is related to marriage practices among Native-Americans.

Part III - Matching I

Match the following descriptions to one of the theoretical models. Obviously, some answers may be used more than once and a few are typical of more than one theory.

- A. Symbolic interaction theory
- B. Social exchange theory
- C. Structural functionalism theory
- D. Conflict theory
- E. Family systems theory
- F. Family development theory

___ 1. Focuses on roles and role playing.

___ 2. Assumes competition for power.

___ 3. Focuses on the study of society rather than individuals.

___ 4. Does not recognize the power of love or bonding.

___ 5. Fears social instability.

___ 6. Emphasizes verbal and non-verbal communication.

___ 7. Assumes that people are most happy when they get what they feel they deserve in a relationship.

___ 8. Views the family as seeking a goal of homeostasis.

___ 9. Views marriage as a series of trade-offs.

___ 10. States that the family has a structure which can only be seen in its interactions.

___ 11. Gauges the values of costs, rewards and resources.

___ 12. Sees boundary maintenance as important.

___ 13. Views conflict as an accepted part of a relationship and as normal.

___ 14. Describes the interacting influences of family roles and circumstances over time.

Matching II: Match the research method with the statement which most characterizes it.

    A. Surveys (questionnaire)   D. Direct observation
    B. Surveys (interview)   E. Experimental research
    C. Clinical studies

__C__ 1. Analyzes individuals and families in a therapy situation.
__A__ 2. Offers anonymity, may be completed fairly quickly, and is relatively inexpensive to administer.
__D__ 3. A researcher unobtrusively studies a family in interaction, without participating.
__B__ 4. A researcher asks questions of an individual in person or by telephone. _interview_
__E__ 5. Researchers can reasonably determine which variables affect the other variables. _experimental_

SELF-DISCOVERY

List each of the members of your family and the kinds of power each uses to affect decisions. Be sure to include yourself.

What are some of the roles that you play within your own family?

Chapter 2

Are there roles that you would like to alter? Which ones? Why is it hard to change roles? How might you go about doing this?

Examine a current significant relationship in your life. (This could be a friendship, romance, spouse, parent, child.) List the benefits and the costs of this relationship.

Relationship _____

	BENEFITS	COSTS
1.		
2.		
3.		
4.		
5.		
6.		
7.		
8.		

Do you think that the costs and benefits balance each other out? If not, does this reflect stress? What can you do to change the balance?

JUST FOR FUN

Imagine that you encountered the following questionnaire in a women's magazine:

SURVEY OF MARITAL HAPPINESS

1. Your age:
   - ____ a. real young
   - ____ b. 20-25
   - ____ c. 25-35
   - ____ d. over 35
   - ____ e. old

2. Your social class:
   - ____ a. real rich
   - ____ b. upper-middle class
   - ____ c. middle class
   - ____ d. common worker
   - ____ e. poor
   - ____ f. welfare recipient

3. Your occupation:
   - ____ a. manager, owner
   - ____ b. sales worker
   - ____ c. clerical
   - ____ d. teacher
   - ____ e. just a housewife
   - ____ f. not for discussion
   - ____ g. mortician
   - ____ h. mom

4. How would you describe your spouse:
   - ____ a. a go-getter
   - ____ b. so-so
   - ____ c. a pinch-hitter
   - ____ d. the light of my life
   - ____ e. who?
   - ____ f. inert

5. How would you describe your love life?
   - ____ a. I'd rather cuddle and forget the rest.
   - ____ b. I'd rather forget the cuddling and go for it!
   - ____ c. I'd rather forget the whole thing.
   - ____ d. I forgot.
   - ____ e. Is that all there is?

As a researcher, would you have any trouble with the choices for each question? If so, what are the problems?

Are the choices equivalent and unbiased?

Continued on next page

Chapter 2

Were the questions worded in an objective manner that would elicit an honest response?

Was the questionnaire written in such a way that it would attract a written response?

What other items might have been asked?

Read the following analysis of the results:

> **THE RESULTS OF OUR SURVEY INDICATE THAT MARRIED PEOPLE WOULD RATHER FORGET THE WHOLE THING!**
>
> With the results of our national survey back, *National Lurid Magazine* has discovered that the love life of Americans is in trouble!
>
> Of the hundred and three responses (97% came from one mid-west town), we found that most of our respondents were middle-class, middle-aged, middle-of-the-road and bored. This representative sample proves the question conclusively!

KEY TO SELF QUIZZES

Multiple Choice		True/False		Matching I		Matching II
1. d	10. d	1. F	11. F	1. a	10. e	1. c
2. b	11. e	2. F	12. T	2. d	11. b	2. a
3. c	12. e	3. F	13. T	3. c	12. e	3. d
4. d	13. b	4. T	14. T	4. d	13. d	4. b
5. e	14. c	5. F	15. T	5. c	14. f	5. e
6. a	15. c	6. T	16. F	6. a		
7. a	16. b	7. F	17. T	7. b		
8. a	17. d	8. T	18. F	8. e		
9. d	18. e	9. T	19. F	9. b		
		10. T	20. F			
			21. T			
			22. T			

SUGGESTED READINGS

For related readings, see page 75 of the text.

CHAPTER 3
Contemporary Gender Roles

MAIN FOCUS

Chapter Three examines the social meanings given to the gender differences between men and women. Gender roles greatly affect how we see ourselves and others, our expectations of marriage and parenting, and our evaluation of our life's successes. These roles have become more androgynous in the last twenty years, resulting in significant changes in both relationships and society.

GOALS OF THIS CHAPTER

To demonstrate mastery of this chapter, you should be able to:

1. Understand and define gender role, gender role stereotype, gender role attitudes, gender role behavior, and gender identity.
2. Discuss the concept of **bipolar gender roles**.
3. Explain how **gender schema** exaggerates male-female differences.
4. Detail the basic assumptions and primary focus of the prominent theories used to explain the significance of gender in our culture: gender theory, social learning theory, and cognitive development theory.
5. Know and describe the socialization agents from birth through adulthood.
6. Describe how children learn their gender roles through manipulation, channeling, verbal appellation, and activity exposure.
7. Describe the traditional roles for men and women, and why those roles may be individually limiting.

Chapter 3

8. Describe contemporary gender roles for men and women, and the ways that these are changing.
9. Discuss the constraints of contemporary gender roles and their resistance to change.
10. Understand and define the concept of androgyny, its important aspects, and ways the concept has been attacked.
11. Discuss the role of gender in music videos.
12. Explain the ramifications of American gender roles for immigrants.

KEY TERMS AND IDEAS

The following terms, ideas, and concepts are listed in the order in which they appear in Chapter Three. Be sure that you understand and can define each of the following:

instrumental trait	gender-role stereotype	gender schema
expressive trait	gender-role attitudes	social construct
gender	gender-role behaviors	modeling
role	gender identity	peers
gender role	bipolar gender role	androgyny

CHAPTER THREE OUTLINE

I. INTRODUCTION TO THE CHAPTER
   A. The traditional view of masculinity and femininity sees men and women as opposites.
   B. According to this view, men have **instrumental traits**, that is, ones which are task-oriented.
   C. The traditional view sees women as having **expressive traits**, which are emotion-oriented traits.

II. UNDERSTANDING GENDER AND GENDER ROLES
   A. Defining key terms for studying gender roles is important to avoid confusion.
      1. **Gender** refers to male or female.
      2. **Role** refers to culturally defined expectations a person is expected to fulfill in a given situation and culture.
      3. **Gender roles** refer to the roles a person is expected to perform as a result of being male or female in a particular culture.
      4. **Gender role stereotypes** refer to rigidly held and oversimplified beliefs that all males and females possess distinct psychological and behavioral traits because of their gender.
      5. **Gender role attitudes** refer to beliefs we have about ourselves and others regarding appropriate male and female personality traits and activities.

6. **Gender role behaviors** refer to the actual activities or behaviors we engage in as males and females.
B. **Gender identity** is based on genitalia, and learned at a very young age.
   1. **Gender identity** is perhaps the deepest concept we hold of ourselves.
   2. Cultures determine the content of gender roles in their own ways.
C. Masculinity and femininity: Opposites or similar?
   1. Until the last generation, the **bipolar gender role** was the dominant model used to explain male-female differences.
      a. According to this model, males and females are polar opposites.
      b. Males possess exclusively instrumental traits.
      c. Females possess exclusively expressive ones.
      d. While sociologists no longer use this model, Americans' beliefs related to gender roles have changed little.
   2. The problem with the view that men and women are opposites is that it is not true. Men and women are more alike than different.
   3. Most differences, where they do exist, can be attributed to gender-role expectations, male-female status, and gender stereotyping.
D. Gender schema is one way culture exaggerates existing gender differences or creates differences where none otherwise exist.
   1. **Gender schema** is a set of interrelated ideas that help us process information by categorizing it in useful ways according to gender.
   2. Adults who have strong gender schema quickly categorize people's behavior and personality characteristics into masculine versus feminine categories.
   3. Many people feel uncomfortable if they don't know a person's gender.

III. GENDER AND SOCIALIZATION THEORIES
   A. Gender theory is a feminist theory begun in the 1970s, and more thoroughly developed in the 1980s, which attempts to explain inequality.
      1. Gender theory is based on two assumptions:
         a. Male-female relationships are characterized by power issues.
         b. Society is constructed in such a way that males dominate females.
      2. According to gender theory, gender is a **social construct**—that is, an idea or concept created by society through the use of social power.
      3. Gender theory focuses on:
         a. How specific behaviors or roles are defined as male or female.
         b. How labor (paid and unpaid) is divided into man's work and woman's work.
         c. How different institutions bestow advantages on men.
      4. The key to the creation of gender inequality is the belief that men and women are "opposite" sexes.

B. Social learning theory, from behaviorist psychology, suggests that we learn attitudes and behaviors as a result of social interaction with others.
1. The cornerstone of social learning theory is the belief that consequences control behavior.
2. Positive reinforcement rewards behavior, while negative reinforcement makes it less likely to recur.
3. The behaviorist approach has been modified to include cognition—mental processes that intervene between stimulus and response, such as evaluation and reflection.
4. A person's ability to anticipate consequences of sex role behavior affects behavior.
5. **Modeling**, learning by imitation, is another way we learn gender roles.
C. Cognitive development theory focuses on the child's active interpretation of messages from the environment.
1. Cognitive development theory stresses the idea that we learn differently depending on our age.
2. Around age six, children recognize that gender is permanent.
3. Children tend to adhere to rigid sex role stereotypes because of an internal need for congruence.

IV. LEARNING GENDER ROLES
A. Gender-role learning in childhood and adolescence is influenced primarily by parents, teachers, peers, and the media.
B. During infancy and early childhood, a child's most important source of learning is the primary caretaker, usually their parent(s).
1. Immediately after birth, parents differentiate in treatment between boys and girls.
2. Children are socialized in gender roles through four processes:
   a. Through manipulation, certain behaviors are reinforced until children accept their parents' views.
   b. Through channeling, children's attention is directed to specific objects.
   c. Through verbal appellation, parents use different words to describe the same behavior by boys or by girls.
   d. Through exposure to different activities or chores.
C. Teachers, as socializing agents, become influential as children enter day care or kindergarten—the child's first experience in the wider world outside the family.
D. Peers, a child's age-mates, become especially important when the child enters school.
1. Peers reinforce gender-role norms through play activity and toys.
2. Peers react with approval or disapproval to other's behavior.
3. Peers influence the adoption of gender-role norms through verbal approval and disapproval.
4. Children's perceptions of their friends' gender-role attitudes, behaviors, and beliefs encourage them to adopt similar ones in order to be accepted.
5. During adolescence, peers continue to have a strong influence, but parents can be more influential than peers.

E. The media, in particular television, tends to condone negative stereotypes about gender, ethnicity, age, and gay men and lesbians.
F. Role transcendence, a life-span perspective to gender-role development, proposes three stages of gender role identity: (1) undifferentiated, (2) polarized, and (3) transcendent.
   1. Gender role learning continues in adulthood and takes place in contexts outside the family of origin.
   2. College encourages young people to think critically and to sometimes consider alternatives to traditional gender roles.
   3. Marriage is an important source of gender role learning, with our partner's expectations shaping our behavior.
   4. Parenthood tends to alter women's lives more than it alters men's lives; when children are born roles tend to become more traditional.
   5. The workplace has different expectations and opportunities for men and for women, creating different attitudes toward achievement.

V. CHANGING GENDER ROLES
   A. Within the past generation, there has been a significant shift from traditional toward more egalitarian gender roles.
      1. Traditional gender-role stereotypes ascribe traits to one gender but not the other, with men showing instrumental traits and women showing expressive traits.
      2. Central features of the traditional male role, regardless of ethnicity, include dominance, work, and family.
         a. Males are generally regarded as more power oriented and demonstrate higher degrees of aggression.
         b. Traditional men see their primary family function as that of provider and are more often confused by their spouse's expectations of intimacy.
      3. There are striking ethnic differences in female roles.
         a. Traditional white female gender roles center around women's roles as wives and mothers.
         b. The traditional female gender role did not extend to African-American women because employment and self-reliance are integral components of their roles of wife and mother.
         c. In traditional Latino gender roles, women subordinated themselves to males out of respect for the male's role as provider.
   B. Contemporary gender roles are evolving from traditionally hierarchical gender roles to more egalitarian and androgynous gender roles.
      1. Women are increasingly taking on the roles of employed workers and professionals, although these may conflict with parenting.
      2. Record numbers of women are choosing not to have children because of the conflicts it creates; this is less true for women from ethnic and minority status groups.
      3. Women have greatly increased their power in decision-making, but husbands continue to have more power in actual practice.

Chapter 3

    4. The mutually exclusive division of traits as either male (instrumental) or female (expressive) is breaking down.
    5. Men are expanding their family roles beyond "breadwinning": Many of those in the evolving Men's Movement share the beliefs of feminism.

VI. CONSTRAINTS OF CONTEMPORARY GENDER ROLES
   A. Although substantially more flexibility is offered to men and women today, contemporary gender roles and expectations continue to limit our potential.
   B. Men are required to work and support their families rather than have the same role freedom to choose to work as women have.
   C. When the man's roles of worker and father come into conflict, it is usually the father role that suffers.
   D. Men continue to have greater difficulty in expressing their feelings and may be out of touch with their inner lives.
   E. Contemporary men still expect, and in many cases are expected, to be dominant in relationships.
   F. Research suggests that the traditional female gender role does not foster self-confidence or mental health: Both men and women tend to see women as less competent then men.
   G. Differences in gender roles have created what Bernard calls the "his" and "her" marriage: Each gender experiences marriage differently.
   H. Due to the "double standard of aging", our culture treats aging men and women differently.
   I. Both men and women reinforce traditional gender role stereotypes among themselves and each other: Despite the limitations of traditional gender roles, changing them is not easy.
   J. The social structure is resistant to change and works to reinforce traditional gender norms and behaviors.
    1. Some religious groups view traditional roles as being divinely ordained.
    2. The marketplace also helps enforce traditional gender roles.

VII. ANDROGYNOUS GENDER ROLES
   A. **Androgyny** refers to the state of combining male and female characteristics.
   B. Androgynous gender roles are characterized by flexibility and a unique combination of instrumental and expressive traits.
   C. Individuals who are rigidly both instrumental and expressive, despite the situation, are not considered androgynous
   D. Androgynous individuals and couples appear to have a greater ability to form and sustain intimate relationships and adopt a wider range of behaviors and values.
   E. Some researchers are not convinced that androgyny leads to greater psychological health and happiness, but rather, that it may contribute to additional sources of inadequacy.
   F. Contemporary gender roles are still in flux: Few men or women are entirely egalitarian or traditional.

VIII. FEATURES AND READINGS
   A. In the *Perspective: Gender Roles and Music Videos*, the influence of the visualization of music on the imaging of gender roles is examined: Music videos present popular images of masculinity and femininity.
   B. *Other Places Other Times: Becoming 'American': Immigrants and Gender Roles in Transition*, examines the problems immigrants face in negotiating the differences in gender roles between their original culture and American culture.
   C. *You and Your Well-Being: Gender and Stress*, discusses the characteristic stresses resulting from contemporary gender roles.
   D. *Understanding Yourself* presents the Bem Sex Role Inventory, one instrument used to measure masculinity, femininity, and androgyny in individuals.

SELF-DISCOVERY I

Examine again your family of orientation. In what ways were your mother (or female role model) traditional? Androgynous?

In what way was your father (or male role model) traditional? Androgynous?

What messages were you given by your family regarding your appropriate gender role behavior? Were these messages more traditional or androgynous?

Chapter 3

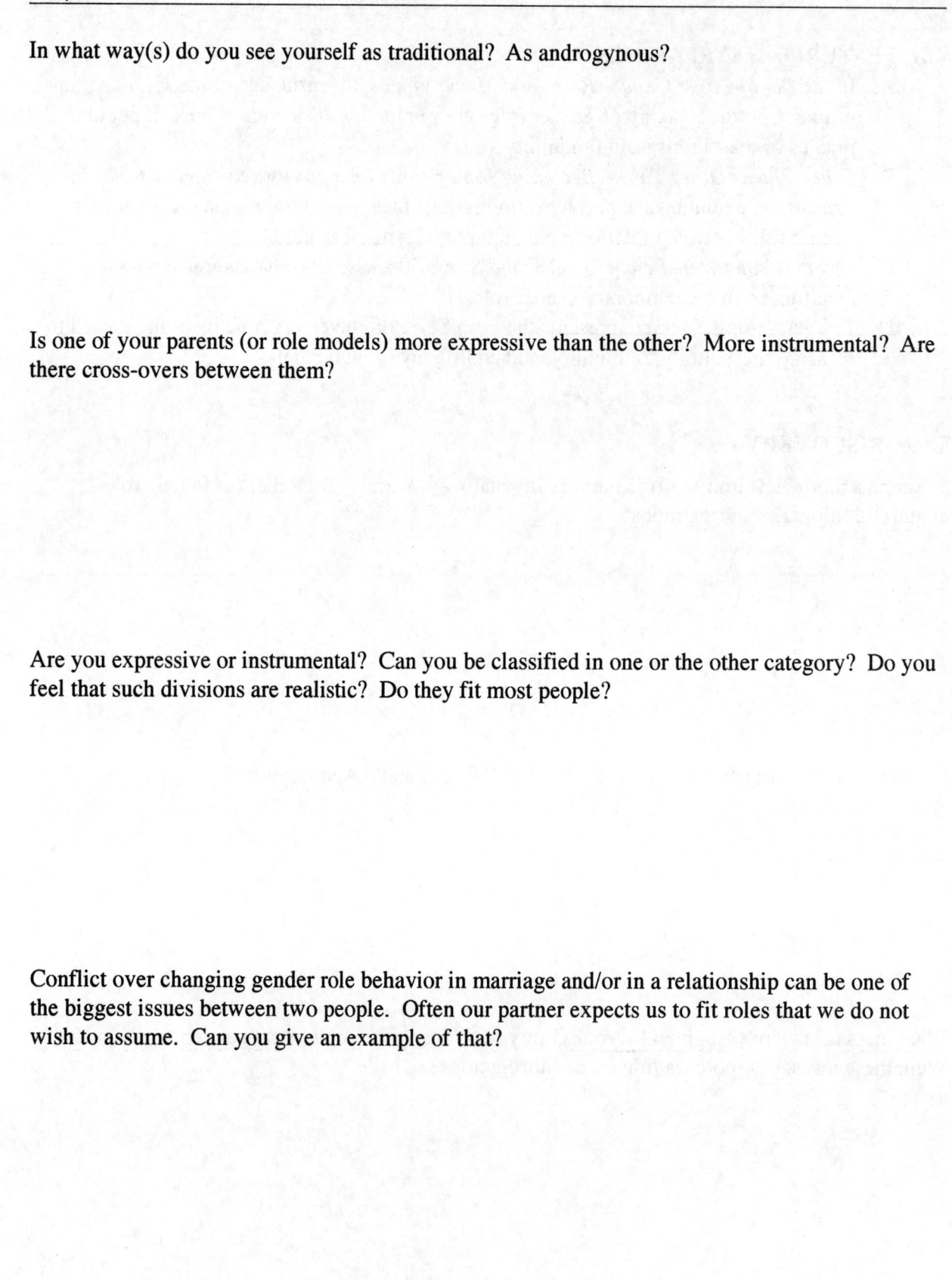

In what way(s) do you see yourself as traditional? As androgynous?

Is one of your parents (or role models) more expressive than the other? More instrumental? Are there cross-overs between them?

Are you expressive or instrumental? Can you be classified in one or the other category? Do you feel that such divisions are realistic? Do they fit most people?

Conflict over changing gender role behavior in marriage and/or in a relationship can be one of the biggest issues between two people. Often our partner expects us to fit roles that we do not wish to assume. Can you give an example of that?

Contemporary Gender Roles

TEST YOUR COMPREHENSION

This chart is to help you examine traditional gender roles in contrast with androgyny. What are the advantages and disadvantages of each system?

Chart 3

ADVANTAGES VS. DISADVANTAGES OF CHANGING GENDER ROLES		
	ADVANTAGES	DISADVANTAGES
Traditional Female Role	wives + mothers	
Traditional Male Role		
Androgynous Female Role		
Androgynous Male Role		

Chapter 3

JUST FOR FUN

The following questionnaire asks you about your beliefs regarding the roles of men and women, husbands and wives. There are no "right" or "wrong" answers, only your opinion. At the end of this review section, you can find out if your responses tend to fall under the "liberal" and androgynous category, or the "conservative" and traditional category. Neither is inherently "good" or "bad." It is important, however, that you and your partner are not too divergent in your attitudes and expectations, or you might be headed for a few heated disagreements!

GENDER ROLE ATTITUDE SURVEY

Please mark each statement with one of the following abbreviations:

SA (Strongly Agree)
A (Agree)
U (Uncertain)
D (Disagree)
SD (Strongly Disagree)

_A_ 1. I feel that raising children is really more the responsibility of the mother than it is the father.

_A_ 2. I think that it is fine for a woman to ask a man that she likes out for a date.

_A_ 3. I think that it looks very unfeminine for a woman to express anger outwardly.

_D_ 4. I would feel threatened if a close male friend expressed deep emotional feelings of sadness to me.

_D_ 5. I feel that it is perfectly all right for a woman to take the sexual initiative.

_A_ 6. If a man and a woman have similar salaries and similar financial obligations, I feel that it is perfectly reasonable that they share dating costs.

_A_ 7. If a girl would rather play with a chemistry set than with her dolls, I would encourage her to do so.

_D_ 8. If a boy would rather play with dolls than with trucks, I would encourage him to do so.

_D_ 9. I would be suspicious of a man who wanted a career as a nurse or as a secretary.

_D_ 10. I feel that religious beliefs dictate that a woman is to be submissive to her husband.

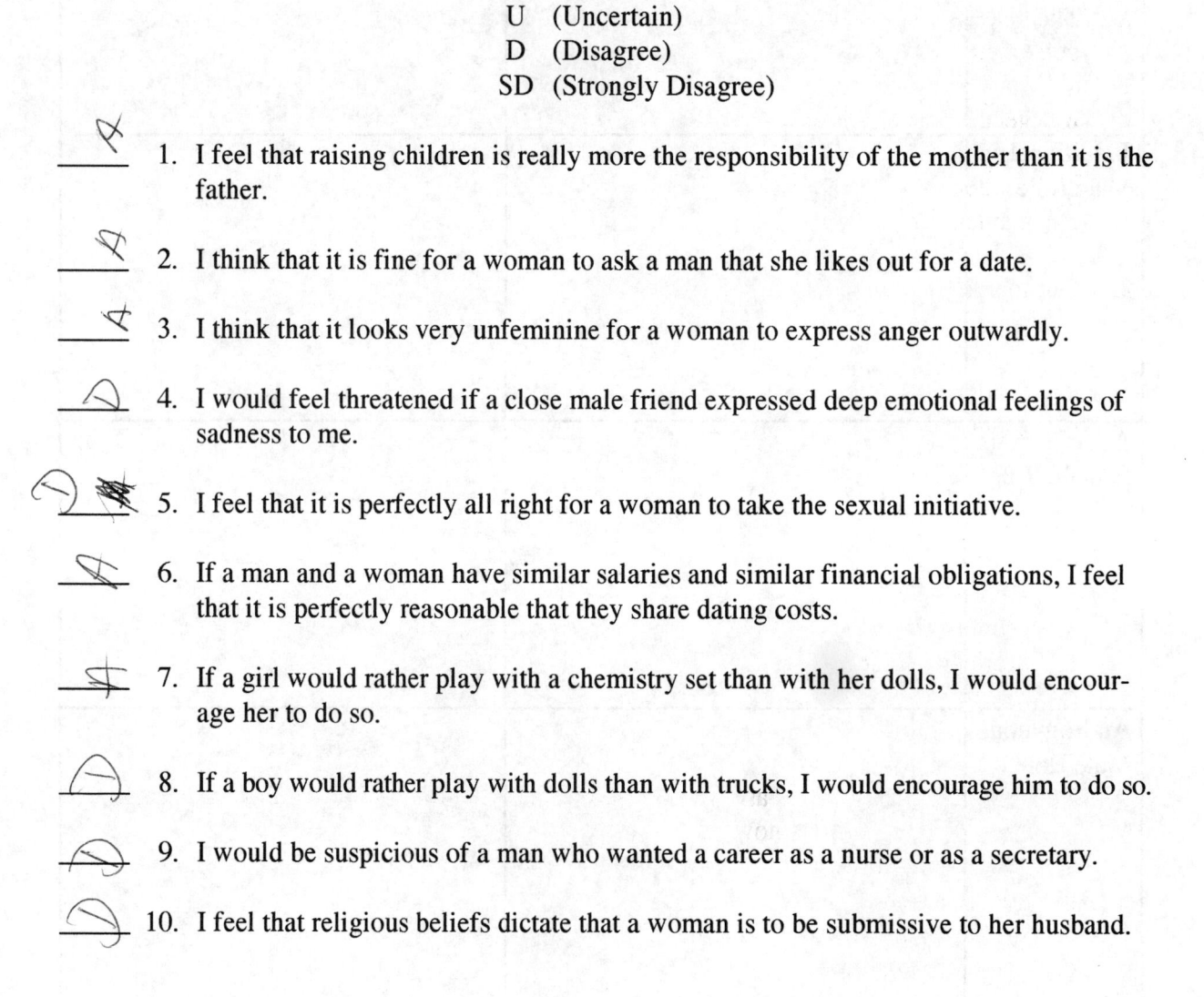

Contemporary Gender Roles

SELF QUIZZES

How well do you know this material? Test your understanding of the reading assignment by answering the following sample questions.

Part I - Multiple Choice: Choose the most correct response.

1. A rigidly held and oversimplified belief that all males and females possess distinct psychological and behavioral traits is called
   a. a gender role.
   b. a gender role stereotype.
   c. gender orientation.
   d. a gender role attitude.
   e. role transcendence.

2. The belief(s) that we have of ourselves and others regarding appropriate male and female personality traits and activities is/are called
   a. gender roles.
   b. gender role stereotypes.
   c. gender orientation.
   d. gender role attitudes.
   e. gender ideology.

3. Our identity which is based on genitalia is called our
   a. gender identity.
   b. gender role.
   c. gender role attitude.
   d. gender role stereotype.
   e. gender difference.

4. Until the most recent generation, the dominant model used to explain male-female differences was
   a. bipolar gender role.
   b. gender schema.
   c. gender modeling.
   d. role transcendence.
   e. none of the above

5. Treating inanimate objects as if they were masculine or feminine is part of a complex structure of associations known as
   a. gender cognition.
   b. gender schema.
   c. gender awareness.
   d. gender recognition.
   e. gender-role attitudes.

Chapter 3

6. Feminist scholars assume
   a. male-female relationships are characterized by power issues.
   b. male dominance is natural.
   c. society is constructed in such a way that males dominate females.
   d. both a and c
   e. all of the above

7. The belief that men and women are "opposite" sexes is the key to the creation of
   a. gender schema.
   b. gender awareness.
   c. gender inequality.
   d. role transcendence.
   e. gender intolerance.

8. Which of the following characterizes social learning theory?
   a. We learn differently depending on our age.
   b. Gender roles are assimilated at different stages of development.
   c. We learn attitudes and behaviors as a result of social interaction with others.
   d. Children understand the permanence of gender but begin to insist on rigid gender role stereotypes.
   e. Children have an internal need for agreement between what they know and how they act.

9. Which of the following characterizes cognitive development theory?
   a. It is derived from behaviorist psychology.
   b. It assumes that positive and negative reinforcement are important learning techniques.
   c. A person's ability to anticipate consequences affects behavior.
   d. Our age and stage of development are directly related to our ability to grasp concepts.
   e. Much learning is done through imitating models.

10. Which of the following is true about gender roles and infants?
    a. Infant boys and girls are held in the same manner.
    b. People who do not know the sex of an infant are extremely uncomfortable in dealing with it.
    c. Baby girls are more likely to be delicate and less noisy than baby boys.
    d. Parents describe boy babies and girl babies in an objective manner.
    e. Baby boys and baby girls behave in different ways almost immediately after they are born.

11. Which of the following characterizes the process of channeling in gender roles?
    a. Manipulating children into appropriate gender role behavior until it becomes an integral part of their personalities.
    b. Directing children to specific toys and objects, based on their gender.
    c. Using different words for boys and girls to describe the same behavior.
    d. Exposing boys and girls to "sex appropriate" activities.
    e. all of the above

12. Peers provide standards for gender-role behavior through
    a. play activities.
    b. non-verbal reactions of approval or disapproval.
    c. verbal approval or disapproval.
    d. toys.
    e. all of the above

13. Which of the following is not true of gender roles in the workplace?
    a. Men and women are psychologically affected by their occupations.
    b. Because men and women have different opportunities for promotion, they have different attitudes towards achievement.
    c. Because many female occupations are low status, women may seem to be less achievement oriented than males.
    d. Although many women choose traditional jobs, there are few instances of sex discrimination in the eighties.
    e. Restrictive jobs tend to lower self esteem and tolerance.

14. Which of the following is not typical of traditional male roles?
    a. The primary function of provider takes precedence over all other family functions.
    b. Traditional men are often confused by their spouses' expectations of intimacy.
    c. Traditional men expect to be the dominant member in a relationship.
    d. They are more likely to be followed by white men than by African American men.
    e. Work is a central feature of the traditional male role.

15. Which of the following is not true of men's roles?
    a. When the man's role of worker comes into conflict with his role as father, usually the latter takes second place.
    b. Men may be out of touch with their emotional selves because they have repressed certain feelings as inappropriate.
    c. Most men recognize the limits of traditional gender roles and would really like to have a totally equal relationship with a woman.
    d. The traditional male role of power makes intimacy difficult to establish.
    e. Male inexpressiveness often makes men strangers to both themselves and their partners.

16. Differences in gender roles are related to all but which one of the following?
    a. "his" and "her" marriage
    b. a double standard of aging
    c. higher satisfaction with marriage for women than men
    d. women being viewed as less competent than men
    e. more wives than husbands desiring divorce

Chapter 3

17. Which of the following does not characterize androgyny?
    a. Masculinity and femininity are seen as mutually exclusive and bipolar.
    b. Androgynous individuals show greater resilience under stress.
    c. Androgynous individuals and couples have a greater ability to form and sustain intimate relationships.
    d. Combining masculine and feminine traits allows us to choose from a full range of emotions and behaviors.
    e. Flexibility and integration are important.

18. All but which of the following are true regarding immigrants and gender roles?
    a. Women tend to experience greater stress than men.
    b. Immigrants must adjust to two gender-role cultures.
    c. Gender-role stress increases for both men and women as their children adopt American gender roles.
    d. Many cultures are far less egalitarian than American culture.
    e. All of the above are true.

Part II - True/False

_____ 1. The perception of male-female differences is far greater than the actual differences themselves.

_____ 2. Our gender identity, based on genitalia, is learned at a very young age and is perhaps the deepest concept we hold of ourselves.

_____ 3. Learning gender roles through imitation is called modeling.

_____ 4. Most parents realize that their words and actions contribute to their children's gender socialization.

_____ 5. Girls generally excel over boys in all areas during grade school.

_____ 6. Parents influence their adolescent's behavior primarily through modeling.

_____ 7. Although women are under greater pressure to marry, studies indicate that marriage contributes more to the happiness of men than it does to the happiness of women.

_____ 8. The traditional female gender role facilitates a sense of self-worth, mental health and self-confidence.

_____ 9. Persons from conservative religious groups adhere most strongly to traditional gender roles.

_____ 10. While gender attitudes within the family have become more liberal, in practice, gender roles continue to favor men in terms of housekeeping and child care activities.

_____ 11. Record numbers of women are rejecting motherhood because of the conflicts it creates with work and time with husbands.

Contemporary Gender Roles

__F__ 12. The public and researchers have expressed increasing social concern about the "working father."

__F__ 13. Men have the same role freedom to chose work as women.

__T__ 14. Studies indicate that unmarried women tend to be happier and better adjusted than married women.

__T__ 15. The traditional view of masculinity and femininity is bipolar.

__F__ 16. Researchers are still convinced that androgyny necessarily leads to greater health and happiness.

__T__ 17. Expectations to be both feminine and masculine may impose stresses equal to those formerly demanded to be either of these.

__T__ 18. Most women in music videos are depicted as sex objects.

DISCUSS BRIEFLY

What is the dominant image of the ideal woman presented in music videos? The ideal man?

Have these images affected you personally or been disturbing to you?

SELF-DISCOVERY II

Are there any incidents in school that come to your mind that were important lessons for you in learning your "appropriate" gender role?

Were there any teachers in your elementary or high school who encouraged androgynous behavior? In what way(s)?

Chapter 3

Have you received any messages from your occupation (or your planned occupation) about gender role expectations?

MINI-ASSIGNMENT

Ask your "significant other" or a good friend to describe an ideal woman and an ideal man. How similar or different are these traits? Were you comfortable with the description of your ideal gender role? Would these descriptions fit the "healthy adult?"

KEY TO "JUST FOR FUN: GENDER ROLE ATTITUDE SURVEY"

Remember, there is no "right" or "wrong" answer, only what you believe! The lower the score (1 - 4), the more conservative and traditional you are likely to be; a higher score, (7 - 10) indicates that you are more likely to be liberal and egalitarian. Give yourself one point for each time your answer agrees.

1. SD, D
2. SA, A
3. SD
4. SD, D
5. SA, A
6. SA, A
7. SA, A
8. SA, A
9. SD, D
10. SD, D

KEY TO SELF QUIZZES

Multiple Choice

1. b
2. d
3. a
4. a
5. b
6. d
7. c
8. c
9. d
10. b
11. b
12. e
13. d
14. d
15. c
16. c
17. a
18. e

True/False

1. T
2. T
3. T
4. F
5. T
6. F
7. T
8. F
9. T
10. T
11. T
12. F
13. F
14. T
15. T
16. F
17. T
18. T

SUGGESTED READINGS

For related readings, see page 103 in the text.

CHAPTER 4

Friendship, Love, & Commitment

MAIN FOCUS

Chapter Four examines friendship, love and commitment, the development of love, approaches to the study of love, unrequited love, jealousy, and the transformation of love.

GOALS OF THIS CHAPTER

To demonstrate mastery of this chapter, you should be able to:

1. Explain the statement "Love is both a feeling and an activity."
2. Describe how friendship, love, and commitment are linked.
3. Discuss the attitudes and behaviors associated with love.
4. Explain the factors that affect commitment.
5. Discuss the similarities and crucial differences between friendship and love.
6. Describe and understand the development of love as explained in the wheel theory.
7. Describe and understand each of the six basic styles of love as formulated by John Lee.
8. Explain intimacy, passion, and commitment in terms of the triangular theory of love.
9. Discuss the theory of love as attachment, the three basic attachment styles, and the similarities between infant love and adult love.
10. Recognize the three different attachment styles underlying the experience of unrequited love.
11. Explain the importance of understanding jealousy and the psychological dimensions of jealousy.

Chapter 4

12. Describe the transformation of love from passion to intimacy and the ways that the disappearance of romance can lead to a crisis in a relationship.
13. Discuss the importance of intimate love in terms of commitment, caring, and self-disclosure.
14. Explain the ways in which love affects sexuality including gender and sexual orientation differences.

KEY TERMS AND IDEAS

The following terms, ideas, and concepts are listed in the order in which they appear in Chapter Four. Be sure that you understand and can define each of the following:

prototypes	mania	unrequited love
wheel theory of love	agape	jealousy
eros	pragma	suspicious jealousy
ludus	triangular theory of love	reactive jealousy
storge	attachment theory	

CHAPTER FOUR OUTLINE

I. INTRODUCTION TO THE CHAPTER
   A. Love is essential to our lives: Love binds us together as partners.
   B. Love is both a feeling and an activity.
   C. The paradox of love is that it encompasses opposites.
   D. Understanding how love works in the day-to-day world may help us keep our love vital and growing.

II. FRIENDSHIP, LOVE, AND COMMITMENT
   A. Friendship, love and commitment are closely linked in our intimate relationships.
      1. Friendship is the foundation for love and commitment.
      2. Love reflects the positive factors, such as caring and attraction, that draw people together and sustain them in a relationship.
      3. Commitment reflects the stable factors (love, obligations, social pressure) that help maintain the relationship for better or worse.
   B. Although love and commitment are related, they are not necessarily connected; one can exist without the other.
   C. Although we may not have a formal definition of love, we do have prototypes (models) of what we mean by love.
   D. Fehr has identified twelve central attributes of love which act as true barometers in relationships.

E. Research on love has found a number of attitudes, feelings, and behaviors associated with love.
 1. Rubin found that caring, needing, trusting, and tolerating the other were identified with love.
 2. Swensen found that romantic love was expressed by: verbally expressing affection; self-disclosing; giving material and nonmaterial evidence; expressing nonverbal feelings in the other's presence; physically expressing love; and tolerating one another.
 3. Research shows that those in love view the world more positively than those not in love.

F. Our commitments seem to be affected by several factors that can strengthen or weaken the relationship.
 1. We have a tendency to look at romantic and marital relationships from a cost-benefit perspective.
 2. Normative inputs for relationships are the values that you and your partner hold about love, relationships, marriage, and family: These values can either sustain or detract from a commitment.
 3. The structural constraints of a relationship will add to or detract from commitment.
 4. Commitments are more likely to endure in marriage than in cohabiting or dating relationships, which tend to be shorter.
 5. Commitments are more likely to last in heterosexual relationships than in gay or lesbian relationships.
 6. Ethnicity may be the greatest predictor of satisfaction and commitment to a friendship.
 7. An enduring marriage is not necessarily a happy marriage.
 8. Because of the overlap between love and commitment, we can mistakenly assume that if someone loves us, he or she is also committed to us.

G. Friendship and love bind us together, provide emotional sustenance, buffer us against stress, and help to preserve our physical and mental well-being.
 1. Todd and Davis found that although love and friendship are alike in many ways, crucial differences make love relationships both more rewarding and more vulnerable.
   a. Best friends were similar to love relationships in several ways: level of acceptance, trust, respect, levels of confiding, understanding, spontaneity, and mutual acceptance.
   b. Lovers had much more fascination and a sense of exclusiveness with their partners than did friends.
 2. Friendship appears to be the foundation for a strong love relationships.
 3. Partners need to communicate and understand the nature of activities and degree of emotional closeness they find acceptable in their spouse's friendships.

Chapter 4

III. THE DEVELOPMENT OF LOVE: THE WHEEL THEORY
   A. Reiss's **wheel theory of love** suggests that love develops and is maintained through four processes: (1) rapport, (2) self-revelation, (3) mutual dependency, and (4) fulfillment of the need for intimacy.
   B. The wheel theory emphasizes interdependence, bi-directionality, and role conceptions.

IV. APPROACHES TO THE STUDY OF LOVE
   A. John Lee described six basic styles of love: **eros**, **mania**, **ludus**, **storge**, **agape**, and **pragma**.
      1. Lee believes that to have a mutually satisfying love affair, a person has to find a partner who shares the same style and definition of love.
      2. The more different two people are in their styles of loving, the less likely it is that they will understand each other's love.
   B. The **triangular theory of love** emphasizes the dynamic quality of love relationships and sees love as composed of intimacy, passion, and decision/commitment.
      1. Intimacy refers to warm, close feelings of bonding in a relationship.
      2. Passion refers to the elements of romance, attraction, and sexuality in a relationship.
      3. The short-term decision/commitment refers to deciding you love someone; the long-term decision/commitment involves the maintenance of love.
      4. Eight ways of classifying love include: liking, romantic love, infatuation, fatuous love, empty love, companionate love, consummate love, and nonlove.
   C. **Attachment theory** maintains that the degree and quality of attachments we experience in early life influence our later relationships.
      1. It examines love as a form of attachment that finds its roots in infancy.
      2. There are three styles of infant attachment: secure, anxious/ambivalent, and avoidant attachment.
      3. In adulthood, infant attachment styles combine with sexual desire and caring behavior to give rise to romantic love.

V. UNREQUITED LOVE
   A. **Unrequited love**, love that is not returned, is a common experience.
   B. According to Arthur Aron and his colleagues, there are three different attachment styles underlying the experience of unrequited love.
      1. The Cyrano style involves the desire to have a romantic relationship with a person regardless of how hopeless the love is.
      2. The Giselle style misperceives the relationship to be more than it really is.
      3. The Don Quixote style involves the general desire to be in love, regardless of whom one loves.

Friendship, Love, and Commitment

VI. JEALOUSY: THE GREEN-EYED MONSTER
   A. Rather than a sign of love, provoking **jealousy** proves nothing except that the other person can be made jealous.
      1. Jealousy may be a more accurate measure of insecurity and possessiveness than love.
      2. Understanding jealousy is important because jealousy is a painful emotion which can make us feel out of control, can cement or destroy a relationship, and is often linked to violence.
   B. Jealousy is an aversive response that occurs due to a partner's real, imagined, or likely involvement with a third person.
   C. Jealousy is a painful experience during which we feel less attractive and acceptable to our partner.
      1. **Suspicious jealousy** occurs when there is either no reason to be suspicious or only ambiguous evidence to suggest a partner is unfaithful.
      2. **Reactive jealousy** occurs when a partner reveals a current, past, or anticipated relationship with another person.
   D. Jealousy acts as a boundary marker by determining how, to what extent, and in what manner others can interact with members of the relationship and vice versa.
   E. Men experience jealousy when they feel their partner is sexually involved with another man: Women experience jealousy over intimate issues.
      1. Both men and women react to jealousy with anger.
      2. Men express anger over jealousy while women tend to suppress it.
   F. Managing jealousy requires the ability to communicate, the recognition by each partner of the feelings and motivations of the other, and a willingness to reciprocate and compromise.

VII. THE TRANSFORMATION OF LOVE: FROM PASSION TO INTIMACY
   A. Passionate love is unstable; romantic love is usually transformed or replaced by a quieter, more enduring love based on intimacy.
      1. Initially, intimacy increases rapidly: As the relationship continues, it decreases and levels off due to drifting apart or latent intimacy.
      2. Passion is subject to habituation; what was once thrilling becomes less so the more we get used to it.
      3. In becoming habituated, we also become dependent; if the person leaves, we experience withdrawal symptoms.
      4. Commitment grows more slowly than intimacy or passion: Our commitments are most affected by how successful our relationship is.
      5. The disappearance or transformation of passionate love is often experienced as a crisis in a relationship.
         a. Intensity of feeling does not necessarily measure depth of love.
         b. The disappearance of passionate love enables individuals to refocus their relationship to include family, friends, and external goals and projects.

6. Romantic love may be highest during the early part of marriage and decline as stresses from childrearing and work intrude on the relationship.
7. In later life, romantic love may play an important role in alleviating the stresses of retirement and illness.
B. Intimate love is enduring; it is based on commitment, caring, and self-disclosure.
1. Commitment is the determination to continue a relationship or marriage; it is based on conscious choice rather than on feelings.
2. Caring is placing another's needs before your own; it entails an I-Thou relationship as opposed to an I-it relationship.
3. Self-disclosure (revealing our hopes, fears, and everyday thoughts) deepens our understanding of each other.
4. Together, commitment, caring, and self-disclosure help transform love, but the most important means of sustaining love is our words and actions.

VIII. FEATURES
A. The *Perspective: Love and Sexuality,* explores the ways in which love affects sex, the various ways which the two are connected or separate, gender differences related to the meaning of sex, sexual orientation, and love
B. *You and Your Well Being: Social Support and Wellness*, discusses the connection between social support and wellness.
C. *Your Style of Love*, based on the work of John Lee, was developed to help men and women identify their style of love.
D. *Other Places ... Other Times: Jealousy in Polygamous Cultures: Tibetan Society*, describes the traditional practice of fraternal polyandry practiced in Tibetan Society.

SELF-DISCOVERY

From your perspective, what is the nature of "real" love or "true" love?

Friendship, Love, and Commitment

To what extent are you and your partner or significant other in agreement on the nature of "real" or "true" love?

What have you learned from your experiences of being in love?

What do you mean when you say "I love you"?

Chapter 4

Ask several people (partner, friends, relatives, acquaintances): "What do you really mean when you say 'I love you'?" How many people had trouble answering this question? Why do you think they had trouble answering it? Were there common themes in the answers? Were your answers similar to those of the people you asked?

Under what circumstances do some people say "I love you" in a manipulative and/or selfish way?

Friendship, Love, and Commitment

TEST YOUR COMPREHENSION

Below is a chart listing individuals discussed in Chapter Four. After each name, identify the person and describe the theory, idea, or quote about love this individual contributed to the chapter.

Chart 4

INDIVIDUALS, THEIR IDENTIFICATION AND THEIR IDEA(S) ON LOVE		
Name	**Identification**	**Idea, Quote, Theory**
Beverly Fehr	what are true barometers in relationships?	12 central attributes of love
Zick Rubin	attitudes, feelings & behaviors associated w/ love	caring, needing, trusting & tolerating is love
Ira Reiss	love develops & is maintained in a 4 processes	(1) Rapport (2) Self-revelation (3) Mutual dependency (4) Fulfillment of the need for intimacy / interdependence — wheel theory of love
John Lee	eros, ludus, mania, storge, agape, pragma	partners share same style & definition of love. The more different their styles of loving, the less likely it is that they will understand each other's love.
Robert Sternberg	love that is not returned — unrequited love	(1) desire to have a romantic love relationship regardless of how hopeless the love is. (2) misperceive the relationship to be more than it really is. (3) Don Quixote style love regardless — 3 different attachment styles
Phillip Shaver		
Tennyson		
Arthur Aron		
Bringle and Buunk		
Hatfield and Walster		
Martin Buber		

65

Chapter 4

SELF QUIZZES

How well do you know this material? Test your understanding of the reading assignment by answering the following sample questions.

Part I - Multiple Choice: Choose the most correct response.

_a_ 1. _____ are models of what we mean by love which are stored in the backs of our minds.
   a. Prototypes
   b. Eros
   c. Paradigms
   d. Revelations
   e. none of the above

_a_ 2. Zick Rubin found that there are four feelings which identify love. Which of the four seems to be the most important?
   a. caring for the other
   b. needing the other
   c. trusting the other
   d. tolerating the other
   e. Rubin found that all four are equally important.

_b_ 3. According to Ira Reiss, factors important in commitment to a relationship include all but which one of the following?
   a. normative inputs
   b. absence of jealousy
   c. structural constraints
   d. balance of costs to benefits
   e. All of the above are important.

_c_ 4. Todd and Davis found that best friends differ from lovers in their levels of
   a. confiding.
   b. acceptance.
   c. fascination.
   d. respect.
   e. trust.

_e_ 5. Which process described in Reiss's wheel of love is influenced by role conceptions?
   a. rapport
   b. self-revelation
   c. mutual dependency
   d. intimacy needs fulfillment
   e. all of the above

66

6. A manic lover is one who is best characterized as [d]
   a. one who plays games in loving.
   b. one who lets his/her love develop over a long period of time.
   c. one who is a very romantic lover with an ideal image in mind of the beloved.
   d. one who is often obsessive, moody, and jealous.
   e. all of the above

7. Which of the following does not characterize the storgic lover? [c]
   a. deep feelings
   b. love between companions
   c. obsessiveness
   d. relationship often develops over time
   e. all of the above

8. According to Sternberg, the type of love which involves intimacy, passion, and commitment is known as [c]
   a. fatuous love.
   b. infatuation.
   c. consummate love.
   d. romantic love.
   e. complete love.

9. According to attachment theory, [e]
   a. experiences in infancy impact adult relationships.
   b. three styles of infant attachment can be seen in adult attachment.
   c. attachment styles reflect one's comfort with closeness.
   d. the likelihood of being optimistic about love relationships is related to attachment.
   e. all of the above

10. Which of the following statements is true with respect to unrequited love? [e]
    a. Unrequited love is a common experience.
    b. People with anxious/ambivalent attachment styles are most likely to experience unrequited love.
    c. Some people experience unrequited love because they have a general desire to be in love with anybody.
    d. Some people experience unrequited love because they wish to have a romantic relationship with a specific person regardless of how hopeless the love is.
    e. all of the above

11. Jealousy [c]
    a. is an instinctive response.
    b. is a necessary ingredient of a deep and loving relationship.
    c. expresses insecurity and dependency.
    d. is found only in monogamous relationships.
    e. is inevitable in any relationship.

Chapter 4

___a___ 12. Jealousy which occurs when a partner reveals a current, past, or anticipated relationship with another person is called
   a. reactive jealousy.
   b. legitimate jealousy.
   c. befitting jealousy.
   d. suspicious jealousy.
   e. none of the above

___c___ 13. Which of the following does not necessarily diminish over time?
   a. intimacy
   b. passion
   c. commitment
   d. both a and c
   e. All of the above will diminish over time.

___b___ 14. Martin Buber's "I-Thou" relationship is related to
   a. self disclosure.
   b. caring.
   c. commitment.
   d. passion.
   e. intimacy.

___e___ 15. Social support
   a. reduces depression.
   b. increases survival ability.
   c. provides secondary prevention.
   d. maintains both physical and mental health.
   e. all of the above

Part II - True/False

___T___ 1. Love is both a feeling and an activity.

___F___ 2. Love and commitment are necessarily connected.

___F___ 3. Love alone is sufficient to make a commitment last.

___F___ 4. Violations of the peripheral features of love and commitment are considered to be more serious than violations of central ones.

___T___ 5. Persons in love view the world more positively than those who are not in love.

___T___ 6. The structure of a relationship influences the commitment to it.

___F___ 7. Ethnicity has minimal influence on satisfaction and commitment to a friendship.

68

Friendship, Love, and Commitment

__T__ 8. For most people, love seems to include commitment and commitment seems to include love.

__T__ 9. Love relationships are both more rewarding and more vulnerable than friendships.

__F__ 10. Friends are different than lovers in their level of confiding in the relationship.

__T__ 11. According to John Lee, to have a mutually satisfying love affair, a person has to find a partner who shares the same style and definition of love.

__F__ 12. Infatuation is usually reciprocated at a similar level between partners.

__T__ 13. Romantic love and infant/caregiver attachment involve similar emotional dynamics.

__T__ 14. Adults characterized as anxious/ambivalent are more likely to be obsessive, jealous, passionate, and experience highs and lows in their relationships.

__F__ 15. Unrequited love is a powerful but rare experience.

__T__ 16. Provoking jealousy proves nothing more than the fact that the other person can be made jealous.

__F__ 17. Jealousy is always irrational.

__T__ 18. Even though passion is subject to habituation, when a person leaves, we may still experience withdrawal symptoms.

__F__ 19. Time tends to diminish and erode commitments as it diminishes intimacy and passion.

__T__ 20. Some studies suggest that while it is true that romantic love may be highest during the earliest part of a marriage, it begins to renew when the children leave home.

__T__ 21. In an "I-Thou" relationship, each person is valued for their humanity and uniqueness.

__F__ 22. Love is more important to heterosexuals than gay men or lesbians.

__T__ 23. Social support facilitates recovery from illness.

Chapter 4

Part III - Matching

Match the style of love and styles of attachment with the description which most characterizes it.

    a. Eros   d. Storge   g. Secure
    b. Mania   e. Agape   h. Anxious/Ambivalent
    c. Ludus   f. Pragma   i. Avoidant

___ 1. Obsessive love, marked by desire for union, high degrees of sexual attraction, jealousy, and emotional highs and lows

___ 2. Often roller-coaster love with nights of sleeplessness and days of pain and anxiety

___ 3. Undemanding, patient love with no expectations

___ 4. Fears intimacy and experiences emotional highs and lows and jealousy

___ 5. Less likely to believe in media images of love and believes that romantic love can last

___ 6. Delights in the tactile, the sensual, the immediate; this love burns brightly but soon flickers and dies.

___ 7. Looks for compatible background and interests

___ 8. Usually begins as friendship and then gradually deepens into love

___ 9. Plays at love; encounters are casual, carefree, and often careless

KEY TO SELF QUIZZES

Multiple Choice

1. a   9. e
2. a   10. e
3. b   11. c
4. c   12. a
5. e   13. c
6. d   14. b
7. c   15. e
8. c

True/False

1. T   9. T   17. F
2. F   10. F   18. T
3. F   11. T   19. F
4. F   12. F   20. T
5. T   13. T   21. T
6. T   14. T   22. F
7. F   15. F   23. T
8. T   16. T

Matching

1. h
2. b
3. e
4. i
5. g
6. a
7. f
8. d
9. c

SUGGESTED READINGS

For related readings, see page 133 of the text.

CHAPTER 5

Communication and Conflict Resolution

MAIN FOCUS

Chapter Five examines communication patterns and marriage, nonverbal communication, developing communication skills and self-disclosure. It also focuses on the role of conflict within intimate relationships, types of conflicts and ways of resolving them.

GOALS OF THIS CHAPTER

To demonstrate mastery of this chapter, you should be able to:

1. Understand the impact of communication patterns both before and in marriage, the characteristics of communication patterns in the satisfied marriage, and husband/wife differences in marital communication.

2. Describe the role of nonverbal communication and its importance as a means of communication, and examine the relationship between nonverbal and verbal communication.

3. Explain the importance of proximity, eye contact, and touch in nonverbal communications.

4. Discuss the importance of developing communication skills and be able to describe techniques for improving communications.

5. Discuss the four styles of miscommunication.

6. Describe the importance of self-disclosure, trust, feedback, and mutual affirmation in an intimate relationship and describe ways in which these skills can be acquired.

7. Explain the relationship between conflict and intimacy.

Chapter 5

8. Explain the different types of conflict that can exist in a relationship and evaluate methods of conflict resolution.

9. Understand types of power and how power relationships may be changing.

10. Examine the emotion of anger and ways of coping with it.

11. Discuss the relationship between communication traits, conflict resolution, and marital satisfaction.

12. Explain how conflict can be solved through negotiation.

KEY TERMS AND IDEAS

The following terms, ideas, and concepts are listed in the order in which they appear in Chapter Five. Be sure that you understand and can define each of the following:

self-disclosure	feedback	hierarchy of rules
honeymoon effect	power	family rules
proximity	relative love and need theory	meta-rules
trust	principle of least interest	

CHAPTER FIVE OUTLINE

I. INTRODUCTION TO CHAPTER
   A. Intimacy and communication are inextricably connected.
   B. Communication for its own sake involves the pleasure of being in each other's company, the excitement of conversation, the exchange of touches and smiles, and loving silence.
   C. Communication patterns are strongly associated with marital satisfaction.

II. COMMUNICATION PATTERNS AND MARRIAGE
   A. Premarital communication patterns are related to marital satisfaction.
      1. How well a couple communicates before marriage can be an important predictor of later marital satisfaction.
      2. **Self-disclosure**, the revelation of deeply personal information about one's self, prior to marriage is related to relationship satisfaction later.
      3. A couple's negative or positive communication pattern has little effect on marital satisfaction the first year of marriage—this quality is known as the **honeymoon effect**.
   B. Couples in satisfied marriages tend to have these marital communication patterns:
      1. Willingness to accept conflict but to engage in it in nondestructive ways.
      2. Less frequent conflict and less time spent in conflict.

3. The ability to disclose or reveal private thoughts and feelings, especially positive ones, to one's partner.
4. Expression by both partners of more or less equal levels of affection.
5. More time spent talking, discussing personal topics, and expressing feelings in positive ways.
6. The ability to accurately encode and decode verbal and nonverbal communication.

C. Gender differences in partner communication are influenced by gender differences in general communication patterns.
1. Wives send clearer messages to their husbands than vice-versa and tend to be more sensitive and responsive to their husbands' messages, both during conversation and conflict.
2. Husbands tend to give more neutral messages or to withdraw.
3. Although communication differences in arguments between husbands and wives are usually small, they nevertheless follow a typical pattern with wives tending to set the emotional tone of the argument.
4. Studies suggest that poor communication skills precede the outset of marital problems.

III. NONVERBAL COMMUNICATION
A. There is no such thing as not communicating.
1. Even when we are not talking, we communicate through silence or gestures.
2. One of the problems with nonverbal communication is the imprecision of its message.

B. The functions of nonverbal communication include conveying interpersonal attitudes, expressing emotions and handling the ongoing interaction.

C. A relationship exists between verbal (basic content) and nonverbal (relationship) messages: For a message to be most effective, these two components must be in agreement.

D. Three of the most important forms of nonverbal communication are proximity, eye contact, and touch.
1. Nearness in terms of physical space, time, and so on, is referred to as **proximity**.
2. Making eye contact with another person, if only for a second, is a signal of interest.
3. Touch is the most basic of all senses: It is extremely important in human development, health and sexuality.
   a. Many of our words for emotions are words referring to physical contact: "attraction","attachment","feeling".
   b. Touch can also be a violation: Sexual harassment includes unwelcome touching.
   c. Touching often signals intimacy, immediacy, and emotional closeness: Touching goes "hand in hand" with self-disclosure.
   d. Levels of touching differ between cultures and ethnic groups.
   e. Sexual behavior is skin contact: In sexual interactions, touch takes precedence over sight.
   f. Although the value of nonverbal expression may vary between groups and cultures, the ability to communicate and understand nonverbally remains important in all cultures.

Chapter 5

IV. DEVELOPING COMMUNICATION SKILLS
  A. Virginia Satire suggests that people use four styles of miscommunication.
     1. Placaters are always agreeable.
     2. Blamers act superior.
     3. Computers are very correct and reasonable.
     4. Distractors act frenetic and seldom say anything relevant.
  B. We can learn to communicate but it is not always easy.
     1. Traditional sex roles encourage men to be strong and silent, to talk about things instead of feelings.
     2. Personal reasons such as inadequacy, vulnerability, or guilt may restrict communication.
     3. Fear of conflict due to expressing real feelings and desires may lead to their suppression.
  C. Before we can communicate with others we must first know how we feel.
     1. Feelings serve as valuable guides for action.
     2. We often place obstacles to self-awareness by suppressing "unacceptable" feelings until we don't experience them. We may also deny or project our feelings.
     3. Becoming aware of ourselves involves becoming aware of our feelings.
     4. Feelings need to be felt; they do not necessarily need to be acted out or expressed.
  D. Communication which reveals ourselves to others is self-disclosure, an important aspect to intimacy.
     1. In the process of revealing ourselves to others, we discover who we ourselves are and self-disclosure is often reciprocal.
     2. A review of studies on the relationship between communication and marital satisfaction finds that a linear model of communication is more closely related to marital satisfaction than the too-little/too-much curvilinear model: In the linear model, the greater the self-disclosure, the greater the marital satisfaction.
  E. **Trust** is the belief in the reliability and integrity of a person.
     1. Three conditions must be met for trust to develop:
        a. a relationship must exist and have the likelihood of continuing;
        b. we must be able to predict how the other person will likely behave; and
        c. the other person must have other acceptable options available to him or her.
     2. Trust is important in close relationships because it is vital to self-disclosure, and it influences the way in which ambiguous or unexpected messages are interpreted.
  F. Giving **feedback**, the ongoing process in which participants and their messages create a given result and are subsequently modified by the result, is a critical element in communication.
     1. We can react to self-disclosure by remaining silent, venting anger, expressing indifference or by acknowledging our partner's feelings as valid, then responding to his or her statement.

2. We can engage in dialogue and feedback by:
   a. focusing on "I" statements and avoiding "you" statements,
   b. focusing on behavior rather than the person,
   c. focusing feedback on observations rather than inferences or judgments,
   d. focusing feedback on observations based on a more-or-less continuum,
   e. sharing ideas or offering alternatives rather than giving advice,
   f. focusing feedback in terms of its value to the recipient,
   g. focusing feedback on the amount the recipient can process, and
   h. offering feedback at an appropriate time and place.
G. Mutual affirmation, along with self-awareness, self-disclosure, trust and feedback, are essential to communication in close relationships.
   1. Mutual affirmation is made up of three elements: mutual acceptance; liking each other; and expressing liking in both words and actions.
   2. Mutual affirmation entails people telling their partners that they like them for who they are and, that they appreciate the little things as well as the big things that they do.

V. CONFLICT AND INTIMACY
  A. The more intimate two people become, the more likely they may be to experience conflict: It is not conflict itself that is dangerous to intimate relationships; it is the manner in which the conflict is handled.
  B. Conflict is natural in intimacy and does not necessarily represent a crisis in the relationship.
  C. Two types of conflict affect the stability of a relationship.
     1. Conflicts may be basic or nonbasic.
        a. Basic conflicts challenge the fundamental assumptions of rules of a relationship, and may offer no room for compromise.
        b. Non-basic conflicts are disagreements that do not strike at the heart of a relationship and resolution is possible.
     2. Conflicts may occur because of a situation or because of the personalities of the partners.
        a. Situational conflicts or realistic conflicts occur because of a need to make changes in a relationship.
        b. Personality conflicts arise because of personality, such as the needs to vent aggression, dominate or overpower: They are not directed toward making changes, but simply toward releasing pent-up tensions.
  D. Power conflicts within families over who decides and does what are both complex and explosive.
     1. **Power** is the ability or potential ability to influence another person or group.
     2. Power aspects are not always obvious in close relationships because of three aspects:
        a. Many of us believe intimate relationships are based on love alone.
        b. The exercise of power is often subtle: Marital power takes many forms.
        c. Power is not constantly exercised unless an issue is important to both people and they have conflicting goals.

Chapter 5

3. Traditional roles supported the subordination of the wife to the husband, but these roles are changing with women working and equalitarian standards emerging.
4. Power may be based on personalities.
5. Women have considerable power in marriage, but may fail to recognize the extent of their power due to cultural norms.
6. Power is not a simple phenomenon: It is a dynamic, multidimensional process.
7. There are six bases of marital power, according to French and Raven: coercive power, reward power, expert power, legitimate power, referent power, and informational power.
8. **Relative love and need theory** explains power in terms of the individuals' involvement and needs in the relationship.
9. The **principle of least interest** describes the fact that the partner with the least interest in continuing a relationship has the most power in it.
10. The feminist critique of family power suggests that we need to think in terms of the effects of the social structure and marital power, power and critical decision making, and power in a family context.
11. Power vs. intimacy may reflect mutually exclusive traits: For genuine intimacy to exist, there must be equality in the power relationships.

VI. CONFLICT RESOLUTION
  A. Differences and conflicts are part of any healthy relationship and can help solidify relationships.
  B. Dealing with anger takes skill and sensitivity and may require negotiation.
   1. Differences between people may lead to anger, which transforms differences into fights and creates tension, distrust, division and fear.
   2. Most people have learned to handle anger by either venting or suppressing it.
   3. Many couples experience a love/anger cycle involving anger at the point a couple become most intimate with each other.
   4. Suppressed anger ultimately leads to resentment and low-level hostility.
   5. Anger can be recognized as a symptom of something that needs to be changed, leading to negotiation.
  C. The way in which a couple deals with conflict resolution both reflects and contributes to their marital happiness.
   1. Happily married couples display distinctive communication behaviors including:
      a. summarizing of what the other person says into his or her own words,
      b. paraphrasing to put what the other person says into one's own words,
      c. validating the other's feelings and,
      d. clarifying the communication if there is uncertainty.
   2. Unhappy couples display the following patterns:
      a. confrontation rather than trying to understand,
      b. confrontation and defensiveness as alternating patterns, and
      c. complaining and defensiveness as alternating patterns.

D. Fighting about sex involves issues that are sexual as well as using sex as a scapegoat for underlying issues which are unresolved.
E. Couples disagree or fight over money for a number of reasons, one of the most important being power.
   1. Money issues tend to support male dominance.
   2. Financial priorities are a major source of disagreement.
   3. Talking about money is often taboo, although our society is obsessed with money.
F. Resolving conflicts can be accomplished by giving in, imposing your will through power, force, or threat of force, and through negotiation.
   1. In negotiations, both partners sit down and work out their differences until they can come to a mutually acceptable agreement.
   2. There are three major ways conflict can be resolved through negotiation:
      a. Agreement as a gift occurs when a person agrees without coercion, threats, or resentment: It is a gift of love.
      b. Bargaining involves making compromises, seeking the most equitable deal for each partner.
      c. Co-existence involves living with the differences without undermining the basic ties.
G. Communication is the basis for good relationships.
H. Communication and intimacy are reciprocal.

VII. READINGS AND FEATURES
A. The *Perspective: Ethnicity and Communication* compares communication patterns among African Americans, Latinos and Asian-Americans.
B. The *Perspective: Family Rules and Communication* offers a family system analysis of rules, meta-rules, feedback and changing rules within families.
   1. A **hierarchy of rules** is the ranking of rules in order of significance.
   2. **Family rules** are the combined members' rules.
   3. **Meta-rules** are general and abstract, and are superior to family rules.
C. The *Perspective: Family Types and Communication* defines three family types:
   1. Closed-type families which emphasize obligations, conformity to tradition and stability.
   2. Open-type families emphasize consent in opinions and feelings in running the family: They seek intimacy and nurturance but not at the expense of the individual's identity.
   3. Random-type families emphasize individual expression: Their highest value is for each member to exercise his or her freedom unfettered by either tradition or consent.
D. The *Perspective: Ten Rules for Avoiding Intimacy* offers a satirical listing of ways to avoid close relationships.
E. *You and Your Well Being: Getting Psychological Help* discusses issues related to resolving problems by accepting the professional assistance of mental health workers.

Chapter 5

TEST YOUR COMPREHENSION

The following chart illustrates the six bases of marital power. Define each of the six bases, and then give an example of each.

Chart 5

SIX BASES OF MARITAL POWER BY DEFINITION AND EXAMPLE		
TYPE OF POWER	DEFINITION	EXAMPLE
Coercive Power		
Reward Power		
Expert Power		
Legitimate Power		
Referent Power		
Informational Power		

Communication and Conflict Resolution

SELF QUIZZES

How well do you know this material? Test your understanding of the reading assignments by answering the following sample questions.

Part I - Multiple Choice: Choose the most correct response.

__d__ 1. Which of the following is not typical of couples in satisfied marriages?
   a. They share a willingness to accept conflict in nondestructive ways.
   b. They are willing to discuss personal topics.
   c. Both partners express more or less equal levels of affective disclosure.
   d. The couple discloses mostly negative thoughts to each other.
   e. The couple can accurately encode and decode verbal and nonverbal messages from each other.

__b__ 2. The honeymoon effect, as related to a couple's communication pattern, lasts about how long?
   a. 6 months
   b. 12 months
   c. 18 months
   d. 24 months
   e. 36 months

__b__ 3. Which of the following is not true of nonverbal communication?
   a. For a message to be most effective, both verbal and nonverbal components must agree.
   b. Nonverbal messages are almost always precise and clear to the other person.
   c. Our emotional states are often expressed through our bodies.
   d. Nonverbal messages are used to convey interpersonal attitudes.
   e. The nonverbal part of a message contains what is known as the relationship part of a message, while the verbal part contains the basic content.

__b__ 4. All of the following are true regarding touch except
   a. touch is a life-giving force.
   b. studies show that females touch more than males.
   c. touch often signals intimacy.
   d. touching goes "hand in hand" with intimacy.
   e. All of the above are true.

__d__ 5. Partners who are always agreeable and passive use the _____ style of miscommunication.
   a. blaming
   b. distracting
   c. computing
   d. placating
   e. none of the above

79

Chapter 5

___C___ 6. Which of the following is usually not true of self-disclosure in a relationship?
   a. Self-disclosure creates the environment for mutual understanding.
   b. Self-disclosure is reciprocal.
   c. Self-disclosure involves acting out our roles as if they were the source of our deepest selves.
   d. Self-disclosure is a key ingredient to intimacy.
   e. All of the above are true.

___e___ 7. Self-disclosure involves
   a. a degree of trust that this information will not be misused.
   b. a willingness to acknowledge our own vulnerabilities.
   c. an important part of developing a close relationship.
   d. courage to risk rejection and to expose one's self to another.
   e. all of the above

___d___ 8. "I" statements are important because they
   a. assess blame.
   b. tell the partner how he or she should feel.
   c. are likely to make the recipient defensive.
   d. state our own feelings rather than attribute feelings to our partner.
   e. all of the above

___a___ 9. Which of the following is not a positive method for improving communication in a relationship?
   a. agreeing with whatever our partner says, to keep the relationship harmonious
   b. keeping our communication open, using "I" statements
   c. focusing on the behavior rather than the person
   d. focusing feedback on observations rather than inferences or judgments
   e. focusing feedback on offering alternatives rather than giving advice

___c___ 10. There has been general recognition among family experts that
   a. conflict is dangerous.
   b. complete absence of conflict between individuals is an indictor of a strong relationship.
   c. the more intimate a couple becomes, the more likely they may be to experience conflict.
   d. the presence of conflict within a marriage indicates that love is gone.
   e. basic conflicts can usually be easily resolved.

___e___ 11. Which of the following statements best describes why power relationships are often invisible in close relationships?
   a. Intimate relationships are frequently viewed as being based on love alone.
   b. The exercise of power often occurs in subtle ways.
   c. Power is not constantly exercised unless it is an important issue in which there is disagreement.
   d. Marital power takes many forms.
   e. All of the above are important factors.

Communication and Conflict Resolution

__C__ 12. Which of the following is not true of power relationships in a marriage?
   a. The law supports the traditional power relationships of man as authority and wife as subordinate.
   b. Women have considerable power in marriage.
   c. Power rests more on law than on personality within a marriage.
   d. Power relations may be misperceived between husband and wife because they both assume traditional power is held by the man.
   e. Institutional patterns of society give men positions of authority over women.

__C__ 13. According to the principle of least interest, the person in a relationship who is least interested in continuing the relationship
   a. will enjoy the relationship less.
   b. will enjoy the relationship more.
   c. has the most power in the relationship.
   d. is vulnerable to exploitation by the other.
   e. may feel more commitment to the relationship.

__b__ 14. Which of the following is not a communication behavior displayed by happily married couples?
   a. summarizing
   b. confrontation
   c. paraphrasing
   d. validation
   e. clarification

__d__ 15. Which of the following is a positive way to increase intimacy?
   a. Don't talk about anything meaningful.
   b. Never show your feelings.
   c. Always be pleasant.
   d. Be willing to self-disclose.
   e. Never argue.

Part II - True/False

__F__ 1. Communication patterns prior to marriage offer little validity in predicting future marital satisfaction.

__F__ 2. Avoiding or denying that a problem exists is an effective way of minimizing family problems.

__T__ 3. Poor communication skills precede the onset of marital problems.

__F__ 4. Husbands tend to send clearer messages than wives.

Chapter 5

___T___ 5. For a message to be most effective, both the verbal and the nonverbal components must be in agreement.

___T___ 6. Proximity may be misinterpreted due to cultural differences.

___T___ 7. Touch can be so important that without it, the human infant can actually fail to thrive and even die.

___F___ 8. Traditional male sex roles enhance communication in the family.

___F___ 9. Research indicates that the relationship between communication and marital satisfaction fits a curvilinear model.

___T___ 10. The degree to which a person trusts someone influences the way they interpret ambiguous messages.

___F___ 11. If we are upset, it is best to direct our anger at the person rather than at the behavior.

___T___ 12. It is important to focus feedback at an appropriate time and place.

___T___ 13. A love/anger cycle can result in a couple experiencing conflict whenever they begin to establish intimacy.

___F___ 14. Nonbasic conflicts strike at the heart of a relationship.

___T___ 15. If a wife refuses to move with her husband, she is legally answerable to a charge of desertion.

___T___ 16. Power often shifts from person to person.

___F___ 17. We are generally open about money and discuss it openly.

___False___ 18. Dating relationships often prepare a couple for how they will deal with money in marriage.

___T___ 19. Agreement can be a gift of love and is often later reciprocated.

DISCUSS BRIEFLY

1. What are some of the limitations of nonverbal communication in a relationship? In what ways can nonverbal communication enhance a connection?

2. What are the four styles of miscommunication, according to Virginia Satir? Describe and give an example of each.

    a.

    b.

    c.

    d.

3. What are the three most important forms of nonverbal communication? Define and give an example of each.

    a.

    b.

    c.

Chapter 5

4. Describe the three major ways conflicts can be resolved through negotiations, and give an example of each:

   a.

   b.

   c.

SELF-DISCOVERY

Do you tend to affirm or negate more in your relationships with significant others? What are some of the ways you tend to affirm? Negate?

In your current or most recent significant relationship are/were there any basic conflicts? Can you give an example? Non-basic conflicts?

Give a personal example of situational conflict.

Give a personal example of personality conflict.

Give a personal example of power conflict.

MINI-ASSIGNMENT

Touch is an important aspect of nonverbal communication. Experiment with using touch more than you normally do in a communication with a close friend or significant other. Did you find that this affected the communication in any way? Did you feel closer? Was this difficult for you?

Chapter 5

Think of a time when you disclosed something personal to someone who is important to you. Was this hard for you? What type of response did you receive? What may have made the self-disclosure easier?

Experiment with affirming someone with whom you are in a close relationship. How might you show this person that you accept him/her? How might you express that you like this person in words? In actions? What response did you receive?

KEY TO SELF QUIZZES

Multiple Choice   **True/False**

1. d	10. c	1. F	10. T
2. b	11. e	2. F	11. F
3. b	12. c	3. T	12. T
4. b	13. c	4. F	13. T
5. d	14. b	5. T	14. F
6. c	15. d	6. T	15. T
7. e		7. T	16. T
8. d		8. F	17. F
9. a		9. F	18. F
			19. T

SUGGESTED READINGS

For related readings, see page 167 in the text.

CHAPTER 6

Pairing and Singlehood

MAIN FOCUS

Chapter Six examines the many factors affecting the selection of a partner, romantic relationships, singlehood, and cohabitation.

GOALS OF THIS CHAPTER

To demonstrate mastery of this chapter, you should be able to:

1. Understand the marriage marketplace and how it functions.

2. Articulate the effects of the marriage squeeze and the marriage gradient.

3. Understand the importance of initial impressions, and define and describe the "halo" effect, the "rating and dating game," and "trade-offs."

4. Understand and describe how marriage selection is affected by the field of eligibles, endogamy, exogamy, homogamy, and heterogamy.

5. Understand the beginning of relationships—seeing, meeting, and dating.

6. Discuss power and problems in dating relationships.

7. Explain how relationships end and things to keep in mind regarding breaking up.

8. Explain the rise of singlehood as well as relationships in the singles world.

9. Classify different types of single people, identifying each type.

Chapter 6

10. Separate the myths from the realities of being single.

11. Discuss gay and lesbian singlehood.

12. Discuss the types of cohabiting couples, gay and lesbian cohabitation, cohabiting couples versus those who marry, and the positive and negative aspects of cohabitation.

13. Explain how the African-American ratio of males to females influences relationships.

14. Discuss the role of self-esteem in partner selection.

15. Give a cross cultural perspective on love vs. arranged marriage.

KEY TERMS AND IDEAS

The following terms, ideas, and concepts are listed in the order in which they appear in Chapter Six and in the outline. Be sure that you understand and can define each of the following:

marriage squeeze	exogamy	closed fields
marriage gradient	homogamy	open fields
halo effect	heterogamy	lesbian separatists
field of eligibles	stimulus-value-role theory	domestic partners
endogamy		

CHAPTER SIX OUTLINE

1. INTRODUCTION TO THE CHAPTER
    A. Although theoretically we have free choice to select our partners, in reality our choice is somewhat limited and channeled.
    B. Over the last several decades, many aspects of pairing, such as the legitimacy of pre-marital intercourse and cohabitation, have changed considerably, radically affecting marriage.
    C. Marriage has lost its exclusiveness as the only legitimate institution in which people can have sex and share their everyday lives

II. CHOOSING A PARTNER
    A. The marriage marketplace is really a process which involves bargaining and exchange, as well as love.
        1. Each of us has certain resources that make up our marketability.
        2. The idea of exchange as a basis for choosing marital partners is deeply rooted in marriage customs.
        3. The traditional marital exchange is related to gender roles: (1) Men traditionally exchange their status, economic power, and protector role for women's physical attractiveness and nurturing, childbearing, and housekeeping abilities; (2) Women gain status and economic security in the traditional marital exchange.

4. Although changing gender roles have created changes in bargaining, a woman's bargaining position today is still not as strong as a man's.
5. An important factor affecting the marriage marketplace is the ratio of men to women; the scarcer sex gains bargaining power.
6. The gender imbalance reflected in the ratio of available unmarried women to available unmarried men (a.k.a. the **marriage squeeze**) is influenced by age and ethnicity.
7. The **marriage gradient**, the tendency for men to marry women slightly below them in age, education, etc., puts high-status women at a disadvantage in the marriage marketplace.

B. Physical attractiveness is particularly important during the initial and early stages of relationships.
1. The inference of qualities based on looks is based on the **halo effect**—the assumption that good-looking people possess more desirable social characteristics than unattractive people.
2. In the rating and dating game, people prefer attractive people over unattractive people for three reasons: "aesthetic appeal" or a simple preference for beauty; the halo effect; and the status achieved by dating attractive people.
3. People tend to gravitate toward those who are about as attractive as themselves.
4. People tend to choose those who are their equals in terms of looks, intelligence, education, and so forth.
5. Trade-offs are made when lower-ranked traits are exchanged for higher-ranked traits.
6. Men are more likely than women to care about how their partners look; this may be attributed to the disparity of economic and social power.

C. Mate selection occurs from a field of eligible people that our culture approves of as potential partners.
1. **Endogamy** assures that we share common assumptions, experiences and understandings of the world.
2. The principle of **exogamy** prohibits us from marrying within specific groups of people, such as relatives, and people of the same sex.
3. **Homogamy** encourages us to marry people with traits similar to our own.
   a. The most important elements of homogamy are race and ethnicity, religion, socioeconomic status, age and personality characteristics.
   b. The elements of homogamy strongly influence our choice of sexual partners.
4. The tendency to choose a mate whose personal or group characteristics differ from our own is **heterogamy**.
5. Most marriages are between members of the same race.
6. Most religions oppose interreligious marriage.
7. Most people marry within their own social class because of shared values, tastes, goals, occupations, expectations, and educational levels.
8. Most people marry within their same age group, although the man is often slightly older than the woman.
9. People tend to choose partners who share similar personality characteristics.

Chapter 6

        10. Marriages that are homogamous tend to be more stable than heterogamous marriages.
    D. Homogamy alone does not explain mate selection.
        1. Murstein developed a three-stage theory, the **stimulus-value-role theory**, to explain the development of romantic relationships.
        2. The theory is based on exchange: After evaluation, the exchange appears more or less equitable and the couple advances to the next stage.
        3. In the stimulus stage, each person is attracted to the other.
        4. In the value stage, partners weigh each other's basic values.
        5. In the role stage, each person analyzes the other's behavior and potential as spouse and parent.

III. ROMANTIC RELATIONSHIPS
    A. Researchers are shifting from the traditional emphasis on mate selection toward the study of the formation and development of romantic relationships.
    B. Beginning a relationship involves seeing, meeting and dating.
        1. The settings in which people see each other can influence their chances of meeting each other.
            a. **Closed fields** are characterized by a small number of people who are likely to interact whether they are attracted or not (e.g., a small class or seminar).
            b. **Open fields** are characterized by large numbers of people who do not ordinarily interact with each other (e.g., beaches or bars).
        2. Men are the most likely to initiate a meeting directly, whereas women are more likely to wait for the other person to introduce himself or herself or to be introduced by a friend.
            a. Parties are the most common settings in which young adults meet, followed by classes, work, bars, clubs, sports settings, and events related to hobbies.
            b. Unmarried men and women increasingly rely on personal classified ads, which tend to reflect stereotypical gender roles, to meet others.
            c. The problem of meeting is exacerbated for lesbians and gay men because they cannot necessarily assume that the person they are interested in shares their sexual orientation.
        3. Both men and women contribute, although sometimes differently, to initiating the first date.
            a. Men are likely to be more direct while women are often more indirect.
            b. Although many women believe that the traditional male prerogative for initiating the first date is outdated, they are often reluctant to violate it.
    C. Although power does not appear to be a concern for most people in dating relationships, the fact that men have more power, prestige, and status than women cannot help but influence heterosexual interactions.
    D. A number of problems may be associated with dating.
        1. Divergent gender-role conceptions may complicate dating relationships.
        2. Places to go, communication, and unwanted pressure to engage in sex, are common problems for women.

3. Communication, where to go, and shyness are common problems for men.
4. Large numbers of both men and women have sexual involvements outside dating relationships that are considered exclusive.

E. Breaking up is usually painful: Few relationships end by mutual consent.
1. Breaking up is rarely easy, whether a person is initiating it or responding to it.
2. Persons initiating break-ups may be helped if they: (1) are sure that they want to break up; (2) acknowledge that their partner will be hurt; (3) do not continue seeing their former partner as "friends" for awhile; and (4) don't change their mind.
3. If a partner breaks up with you, keep the following in mind: (1) the pain and loneliness you feel are natural; (2) you are a worthwhile person, with a partner or not; and (3) keep a sense of humor.

V. SINGLEHOOD
A. Each year more and more adult Americans are single: Rates vary by ethnicity.
B. The number of never-married singles is rising due to a number of factors:
1. Delayed marriage; the longer one postpones marriage, the greater the likelihood of never marrying.
2. Expanded lifestyle and employment options for women.
3. Increased rates of divorce and decreased likelihood of remarriage, especially among African-Americans.
4. Increased number of women enrolled in colleges and universities.
5. More liberal social and sexual standards.
6. Uneven ratio of unmarried men to unmarried women.
C. Relationships within the singles world tend to be highly independent and emphasize autonomy and egalitarian roles, making it difficult for some to make commitments.
1. Because they make more money, men may not have a strong incentive to commit, marry, or stay married.
2. New research indicates that being single at mid-life may be more problematic for men than women: Single women appeared to have a better psychological well-being than single men.
D. Our culture is ambivalent in its messages about being single versus being married.
1. Popular images of singleness emphasize freedom and glamour, but single people often yearn for marriage.
2. Married people may feel limited and yearn for new experiences and independence.
E. Never-married singles can be grouped into types: ambivalents, wishfuls, resolveds, and regretfuls.
F. Myths of singleness suggest that singles are dependent on their parents, self-centered, have more money, are happier, and view singleness as a lifetime alternative.
G. In reality, singles don't easily fit into married society, have more time, have more fun, are more likely to be sexual, but are more lonely.

Chapter 6

    H. By the 1960s, some neighborhoods in the largest cities became identified with gay men and lesbians.
        1. In these neighborhoods, men and women are free to express affection openly, experience little discrimination or intolerance, and are more involved in lesbian or gay social and political organizations.
        2. The urban gay male subculture which emerged in the 1970s emphasized sexuality; however, relational sex has become normative among large segments of the gay population due to the HIV/AIDS epidemic.
        3. During the late 1960s and 1970s, **lesbian separatists** rose to prominence, however, by the mid-1980s, the lesbian community underwent a "shift to moderation."
        4. Lesbians tend to value the emotional qualities of relationships more than the sexual components, reflecting their socialization as women.
        5. Being female influences a lesbian more than being gay.

V. COHABITATION
    A. Cohabitation appears to be here to stay: In 1994, 5% of the population of unmarried heterosexual couples lived together.
    B. Different reasons to cohabitate include:
        1. temporary casual convenience,
        2. affectionate dating or going together,
        3. economic advantage or necessity,
        4 trial marriage,
        5. respite from being single,
        6. temporary alternative to marriage, and
        7. permanent alternative to marriage,
    C. Reasons for cohabiting include more liberal attitudes about sex, changes in the meaning of marriage, and delayed marriage.
    D. Cohabitation does not appear to threaten marriage.
    E. The most notable social impact of cohabitation is that it delays the age of marriage for those who live together.
    F. Cohabiting couples are less likely to stay together compared to married couples: Having children somewhat stabilizes the couples.
    G. Domestic partnership laws are increasing the legitimacy of cohabitation.
    H. The relationships of gay men and lesbians have been stereotyped as less committed than heterosexual couples because they cannot legally marry, they may not emphasize sexual exclusiveness, and heterosexuals misperceive love between gay and lesbian couples as less "real" than heterosexual love.
        1. Regardless of sexual orientation, most people want a close, loving relationship with another person.
        2. Heterosexual couples tend to adopt a traditional marriage model. Gay couples tend to have a "best friend" model.

I. Cohabitation differs from marriage in several ways:
   1. Living together tends to be more transitory and it involves different commitments than marriage.
   2. Most married people pool money, while cohabiting couples do not pool money.
   3. In marriage, the husband is expected to support his wife and family: In cohabitating relationships, the man is not expected to support his partner.
   4. Compared to marriage, cohabitation receives less social support, except from peers.
J. While cohabiting couples argue that cohabitation helps prepare them for marriage, such couples are statistically as likely to divorce as those who do not live together before marriage.

VII. READINGS AND FEATURES
   A. In *The African American Male Shortage* the authors discuss the consequences of the shortage of single black men. Sociologist Robert Staples credits the shortage of black males to institutional racism: high infant mortality, premature death, devastating homicide rates, poor health care access, HIV/AIDS, and illegal drugs.
   B. *Computer Dating and Homogamy* contains a computer dating questionnaire distributed by Compatibility Plus, a computer dating service.
   C. *You and Your Well-Being: The Role of Self-Esteem in Partner Selection* discusses how qualities of self-concept influence interpersonal relations.
   D. *Other Places ... Other Times* discusses love vs. arranged marriages in cross-cultural perspective.
      1. Marriage customs vary dramatically across cultures, and marriage means very different things in different cultures.
      2. In traditional societies, marriage was important family business.
      3. If marriage in American culture is based on individualism and love, marriage in Bedouin culture (in northern Egypt) is based on family honor and duty.

Chapter 6

TEST YOUR COMPREHENSION

Below is a chart illustrating the advantages and disadvantages of cohabiting. Complete the chart, listing all of the arguments in favor of or against cohabiting from your text, lectures, and discussions with family and peers. You do not have to agree with all of the arguments or find all aspects reasonable or rational.

Chart 6

ADVANTAGES AND DISADVANTAGES OF COHABITING	
ADVANTAGES OF COHABITING	DISADVANTAGES OF COHABITING

SELF-DISCOVERY I

Examine your feelings toward cohabiting. Do you agree or disagree with it for others? For yourself? What factors or reasons influenced your answer(s)?

SELF QUIZZES

How well do you know this material? Test your understanding of the reading assignment by answering the following sample questions.

Part I - Multiple Choice: Choose the most correct response.

___e___ 1. Which of the following illustrates the marriage marketplace?
   a. dowries
   b. bride price
   c. social class, status, and looks
   d. endogamy
   e. all of the above

___b___ 2. The marriage gradient refers to
   a. the ratio of available unmarried women to available unmarried men.
   b. the tendency for women to marry men of higher status.
   c. a gender imbalance in potential marital partners.
   d. an advantage for high-status women.
   e. the ability of the scarcer sex to "weight the rules."

___a___ 3. _____ is particularly important during the initial meeting and early stages of a relationship.
   a. Physical appearance
   b. Friendliness
   c. Sincerity
   d. Intelligence
   e. Sense of humor

___b___ 4. Assuming a person is kind and intelligent because he or she is very attractive is an example of
   a. homogamy.
   b. the halo effect.
   c. aesthetic appeal.
   d. stimulus-value-role theory.
   e. all of the above

___d___ 5. At a social event, we are most likely to gravitate toward
   a. the most physically attractive person in the room.
   b. someone not too attractive, since he or she is not likely to reject us.
   c. the least attractive person present, as he or she is likely to be grateful for our attention.
   d. a person who is about as attractive as ourselves.
   e. anyone at all, because looks are unimportant.

Chapter 6

___a___ 6. Endogamy means selecting a marriage partner who is
   a. from our own nationality, race, and class.
   b. our opposite.
   c. of a similar personality.
   d. as attractive as we perceive ourselves to be.
   e. outside our own family.

___d___ 7. Marrying a person who is very similar to ourselves would best illustrate
   a. endogamy.
   b. exogamy.
   c. heterogamy.
   d. homogamy.
   e. bigamy.

___c___ 8. Couples that are homogamous may find that they
   a. have significantly different values, attitudes, and behaviors.
   b. are likely to lack approval from parents, relatives and friends.
   c. share many aspects of their lives in common.
   d. are likely to be less conventional.
   e. are less likely to be stable than heterogamous couples.

___c___ 9. The best example of a closed field is
   a. a bar.
   b. a large university campus.
   c. a party.
   d. the beach.
   e. none of the above

___b___ 10. In a study by Kamarovsky, which of the following was not cited as a frequent problem of dating among men?
   a. communication
   b. women's liberation
   c. where to go
   d. shyness
   e. the fear of rejection

___a___ 11. Power in dating relationships
   a. is perceived by both sexes to be about equal.
   b. is less of a concern than in marital relationships.
   c. is a less intense source of conflict than power in marriage.
   d. is based on the ability to date others.
   e. all of the above

Pairing and Singlehood

___e___ 12. Which of the following is true regarding breaking up?
   a. Most relationships end by mutual consent.
   b. Loneliness after the break up means you are still in love.
   c. Ambivalence means you made the wrong decision.
   d. Generally, men fall out of love more quickly.
   e. Breaking up is rarely easy, regardless of who initiates it.

___a___ 13. Which of the following is least likely to be an explanation for why many people remain single?
   a. Many Americans have significant negative attitudes toward marriage.
   b. Many new career options and lifestyles are open to women in addition to marriage.
   c. More women are attending colleges and universities.
   d. There are more liberal social and sexual standards today.
   e. More people are choosing to cohabit.

___a___ 14. A person who is voluntarily single and considers singleness as probably temporary is typed as
   a. an ambivalent.
   b. a wishful.
   c. a resolved.
   d. a regretful.
   e. unattractive and unable to get a mate.

___d___ 15. Which of the following is a myth regarding singles?
   a. Singles are lonely.
   b. Singles have more fun.
   c. Singles don't easily fit into married society.
   d. Singles view singlehood as a lifetime alternative.
   e. All but a are myths regarding singlehood.

___e___ 16. In which of the following ways is there likely to be a basic similarity between a cohabiting couple and a married couple?
   a. the handling of money
   b. the expectation that the woman must work
   c. the stability of the relationship
   d. the level of encouragement from society regarding making sacrifices to save the relationship.
   e. None of the above are similarities between cohabiting and married couples.

Chapter 6

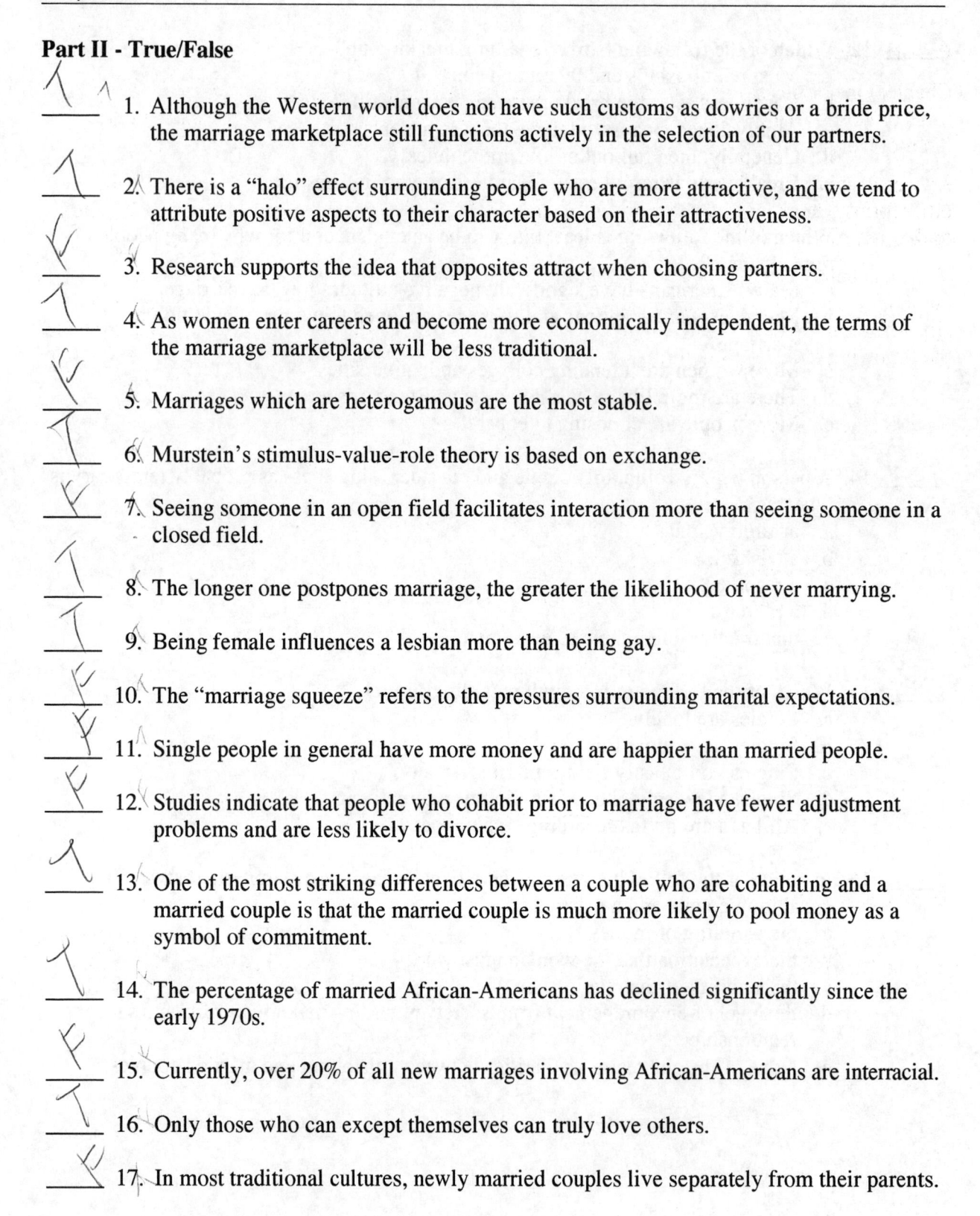

Part II - True/False

___T___ 1. Although the Western world does not have such customs as dowries or a bride price, the marriage marketplace still functions actively in the selection of our partners.

___T___ 2. There is a "halo" effect surrounding people who are more attractive, and we tend to attribute positive aspects to their character based on their attractiveness.

___F___ 3. Research supports the idea that opposites attract when choosing partners.

___T___ 4. As women enter careers and become more economically independent, the terms of the marriage marketplace will be less traditional.

___F___ 5. Marriages which are heterogamous are the most stable.

___T___ 6. Murstein's stimulus-value-role theory is based on exchange.

___F___ 7. Seeing someone in an open field facilitates interaction more than seeing someone in a closed field.

___T___ 8. The longer one postpones marriage, the greater the likelihood of never marrying.

___T___ 9. Being female influences a lesbian more than being gay.

___F___ 10. The "marriage squeeze" refers to the pressures surrounding marital expectations.

___F___ 11. Single people in general have more money and are happier than married people.

___F___ 12. Studies indicate that people who cohabit prior to marriage have fewer adjustment problems and are less likely to divorce.

___T___ 13. One of the most striking differences between a couple who are cohabiting and a married couple is that the married couple is much more likely to pool money as a symbol of commitment.

___T___ 14. The percentage of married African-Americans has declined significantly since the early 1970s.

___F___ 15. Currently, over 20% of all new marriages involving African-Americans are interracial.

___T___ 16. Only those who can except themselves can truly love others.

___F___ 17. In most traditional cultures, newly married couples live separately from their parents.

Pairing and Singlehood

Part II - Matching

Chapter One of this study guide provided examples of learning vocabulary by understanding the roots of words. All of the following terms can easily be remembered if you carefully examine the root word or words.

As was discussed in Chapter One, the root word for marriage is **gamy**, and is combined with other forms (for example, monogamy). **Gyn** is a root for "woman" or "female" (polygyny) and **andro** is a root word for "man" or "male" (polyandry).

New vocabulary components include **endo** meaning "within" (endoderm) and **ex** meaning "without," "outside" (exchange). **Homo** means "same" (homonym) while **hetero** means "opposite" (heteromorphic). **Genos** is "kind," "type" (genotype). Using this etymology, match each of the following terms to its definition:

- a. heterosexual
- b. heterogamy
- c. heterogeneous
- d. homosexual
- e. homogamy
- f. homogeneous
- g. endogamy
- h. exogamy
- i. androgynous

___b___ 1. marrying someone from a dissimilar background
___d___ 2. attracted to the same sex
___g___ 3. marrying within a particular group
___e___ 4. marrying someone of similar type or background
___a___ 5. attracted to the opposite sex
___h___ 6. marrying outside of a particular group
___c___ 7. opposites attract
___f___ 8. likes attract likes
___i___ 9. having both male and female qualities

DISCUSS BRIEFLY

We like to believe that as Americans we're free to marry anyone we wish. For a few of us, this may be true, but for most, society restricts the "field of eligibles." What factors narrow the field and how?

Chapter 6

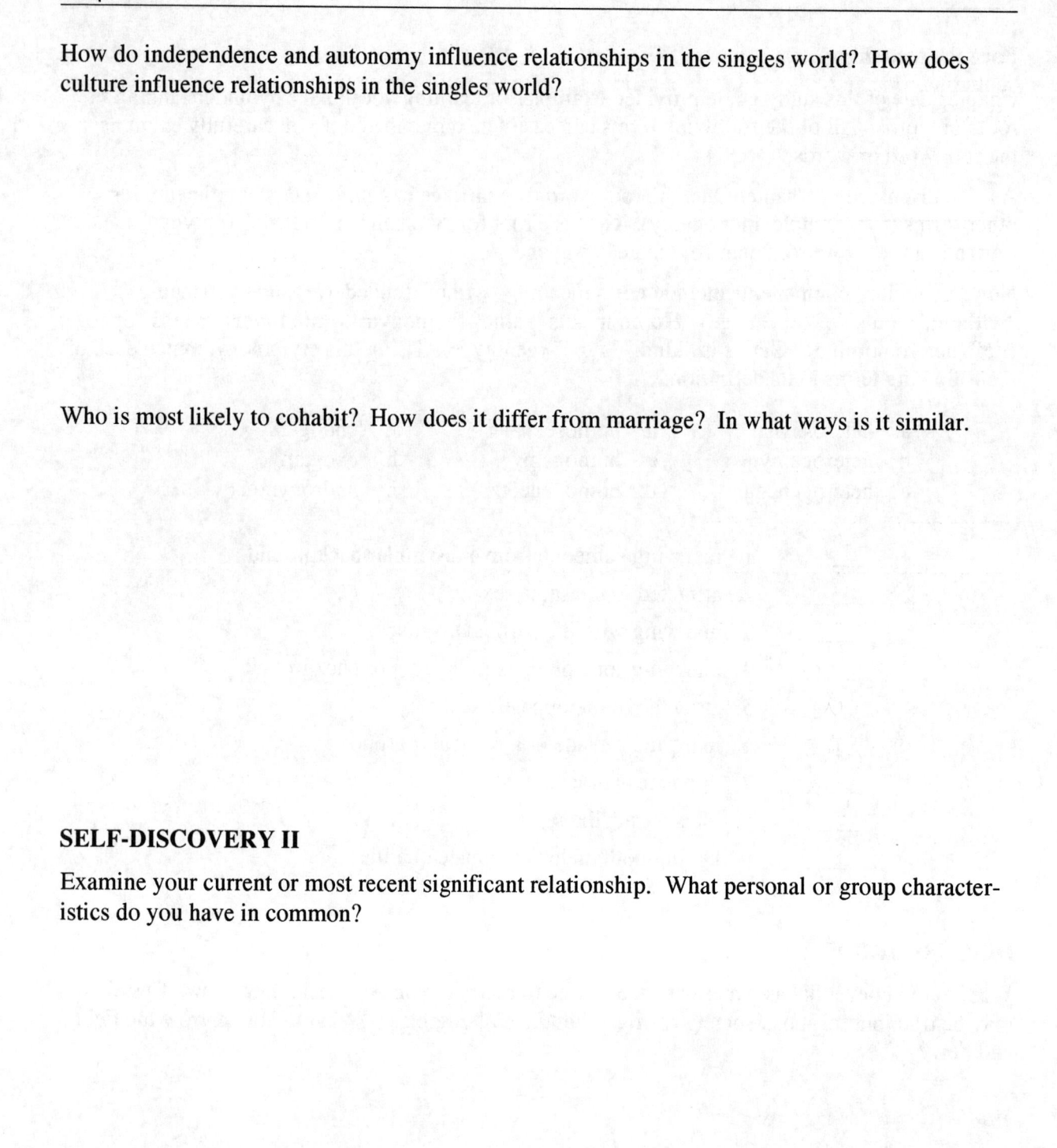

How do independence and autonomy influence relationships in the singles world? How does culture influence relationships in the singles world?

Who is most likely to cohabit? How does it differ from marriage? In what ways is it similar.

SELF-DISCOVERY II

Examine your current or most recent significant relationship. What personal or group characteristics do you have in common?

List the resources (be honest, and it is fine to be positive!) that you bring to this relationship in a column on the left. Now list the resources your partner brings on the right. Do these resources match? Do you see factors of the "marriage gradient" here?

YOU	YOUR PARTNER
_____	_____
_____	_____
_____	_____
_____	_____
_____	_____
_____	_____
_____	_____
_____	_____

MINI-ASSIGNMENT

Examine the "personals" section of one of the local papers. (If one is not available, check the library.) What observations can you make about the kinds of things men are looking for? Women? Are there any universals? Can you make any sociological assumptions about dating and mate selection from this assignment?

If you were to compose your own ad in such a paper, how would it read?

Continued on next page

Chapter 6

Underline what you tell about yourself. Why did you feel this was important?

Why are the characteristics of the person you want to meet important to you?

KEY TO SELF QUIZZES

Multiple Choice   True/False

1. e   10. b   1. T   10. F
2. b   11. e   2. T   11. F
3. a   12. e   3. F   12. F
4. b   13. a   4. T   13. T
5. d   14. a   5. F   14. T
6. a   15. d   6. T   15. F
7. d   16. e   7. F   16. T
8. c   8. T   17. F
9. c   9. T

Matching

1. b   7. c
2. d   8. f
3. g   9. i
4. e
5. a
6. h

SUGGESTED READINGS

For related readings, see page 201 in the text.

CHAPTER 7

Understanding Sexuality

MAIN FOCUS

Chapter Seven examines the sources of sexual learning; psychosexual development in young, middle, and later adulthood; sexual scripts and the gay/lesbian/bisexual identity process; and sexual behavior. It also discusses sexual enhancement; sexual relationships; sexual problems and dysfunctions; birth control; sexually transmitted diseases and HIV/AIDS; and sexual responsibility.

GOALS OF THIS CHAPTER

To demonstrate mastery of this chapter, you should be able to:

1. Describe psychosexual development in young adulthood including: the sources of sexual learning; sexual scripts; developmental tasks; and gay, lesbian, and bisexual identities.

2. Identify developmental tasks and changes related to sexuality in middle adulthood.

3. Identify developmental tasks and changes related to sexuality in older adulthood.

4. Identify and define autoerotic and interpersonal sexual behavior.

5. Identify conditions for good sex and describe means for intensifying erotic pleasure.

6. Compare and contrast premarital, nonmarital, marital, extramarital sexuality, and sexuality in gay and lesbian relationships.

7. Identify the origins of sexual problems and dysfunctions as well as methods for resolving these sexual problems.

8. Detail the issues and factors related to contraception and abortion.

Chapter 7

9. Define and describe the principal STDs and HIV/AIDS, including means of protecting one's self and others.

10. Identify and practice sexual responsibility.

KEY TERMS AND IDEAS

The following columns of terms, ideas, and concepts are listed in the order that they appear in Chapter Seven. Be sure that you understand and can define each of the following:

sexual script	autoeroticism	premarital sex
heterosexual	masturbation	extramarital sex
homosexual	pleasuring	open marriage
bisexual	cunnilingus	sexual dysfunction
gay male	fellatio	abstinence
lesbian	sexual intercourse	contraception
sexual orientation	coitus	abortion
homoeroticism	anal eroticism	Roe v. Wade
coming out	anal intercourse	sexually transmitted disease (STD)
anti-gay prejudice	sexual enhancement	human immunodeficiency virus (HIV)
homophobia	nonmarital sex	acquired immune deficiency syndrome (AIDS)

CHAPTER SEVEN OUTLINE

I. INTRODUCTION TO THE CHAPTER
   A. Being sexual is an essential part of being human.
   B. Paradoxically, sexuality is a source of great pleasure and profound satisfaction, yet it also can be a source of guilt, confusion, infection, and exploitation.

II. PSYCHOSEXUAL DEVELOPMENT IN YOUNG ADULTHOOD
   A. Sources of sexual learning include parents, peers, the media, and partners.
      1. Parental influence—children learn a great deal about sexuality by observing their parents' behavior.
         a. Most parents are ambivalent about their children's developing sexual nature.
         b. Although parental norms and beliefs are generally influential, they do not appear to have a strong effect on an adolescent's decision to become sexually active: Peers appear more important.
      2. Peer influence—adolescents are frequently misinformed about sexuality by their peers.
      3. Media influence—the media and contemporary American popular culture have a profound impact on sexual attitudes.
      4. Partner influence—as we move into young adulthood, sexual partners replace parents, peers, and the media as sources of sexual learning.

Understanding Sexuality

B. Developmental tasks in young adulthood include the following:
1. integrating love and sex,
2. forging intimacy and commitment,
3. making fertility/childbearing decisions,
4. establishing a sexual orientation, and
5. developing a sexual philosophy.

C. A **sexual script** is a set of expectations of how one is to behave sexually as a female or male and as heterosexual, lesbian, or gay; the scripts we are given for sexual behavior tend to be traditional.
1. Female sexual scripts traditionally focus on feelings more than sex. Female sexual scripts include the following ideas:
   a. Sex is both bad and good.
   b. Don't touch me "down there".
   c. Sex is for men.
   d. Men should know what women want.
   e. Women shouldn't talk about sex.
   f. Women should look like models.
   g. Women are nurturers.
   h. There is only one right way to experience orgasm.
2. Male sexual scripts traditionally view sex as performance, and incorporate a separation of sex from love and attachment. Male sexual scripts include:
   a. Men should not have feelings.
   b. Performance is what counts.
   c. The man is in charge.
   d. A man always wants sex and is ready for it.
   e. All physical contact leads to sex.
   f. Sex equals intercourse.
   g. Sexual intercourse always leads to orgasm.
3. Contemporary sexual scripts have an increased recognition of female sexuality, and tend to be more relationship-centered rather than male-centered. Contemporary sexual scripts include the following:
   a. Sexual expression is positive.
   b. Sexual activities are a mutual exchange of erotic pleasure.
   c. Partners are equally involved and responsible.
   d. Masturbation and oral sex are legitimate sexual activities.
   e. Sexual activities can be initiated by either partner.
   f. Both partners have a right to experience orgasm.
   g. Nonmarital sex is acceptable within the context of a relationship.
   h. Gay, lesbian, and bisexual orientations are more open and tolerated.

D. Gay, Lesbian, and Bisexual Identities
1. The term **heterosexual** refers to people who are sexually attracted to members of the other gender; **homosexual** refers to people who are attracted to members of the same gender; and **bisexual** refers to those who are attracted to both genders.
2. The term **gay male**, referring to men, and **lesbian**, in reference to women, replaces the term homosexual, which conveys negative connotations.
3. **Sexual orientation**, that is, a person's sexual identity as heterosexual, gay, lesbian, or bisexual, is a complex interaction of numerous social and personal factors.
4. Identifying oneself as gay or lesbian takes time and includes several stages: **Homoeroticism**, which is the erotic attraction to members of the same gender—generally precedes gay or lesbian activities by several years.
5. The stages in acquiring a lesbian or gay identity usually begin in late childhood or early adolescence. These stages are:
   a. a fear or suspicion that one's desires are somehow different from others,
   b. labeling feelings of love, attraction, desire as homoerotic, and
   c. self-definition as lesbian or gay,
   d. additionally, for some gay men and lesbians, entering the gay subculture and incorporating a way of being in which sexual orientation is a large part of the identity of a person,
   e. and finally, having the first lesbian or gay love affair, and then,
   f. **coming out** by publicly acknowledging one's gayness.
6. **Anti-gay prejudice** is a strong dislike, fear, or hatred of lesbian and gay men.
   a. **Homophobia** is an irrational or phobic fear of gay men or lesbians.
   b. Anti-gay prejudice is derived from a deeply rooted insecurity, a strong fundamentalist religious orientation, and ignorance.
7. Bisexuality requires a rejection of both heterosexual and homosexual identities, and develops in multiple stages.

III. PSYCHOSEXUAL DEVELOPMENT IN MIDDLE ADULTHOOD
A. Developmental tasks in middle adulthood include:
1. redefining sex in marital and other long-term relationships,
2. reevaluating one's sexuality, and
3. accepting the biological aging process.
B. Sexuality and Middle Age
1. As men age, they fear the loss of their sexual capacity but not their attractiveness.
2. As women age, they fear the loss of their attractiveness but not their sexuality.
3. Women reach their peak of sexual responsiveness in the late thirties or early forties; it is usually maintained at about the same level into the sixties and beyond.
4. Men reach their peak of sexual responsiveness in late adolescence or early adulthood; however, changes in sexual responsiveness do not become apparent until men are in their forties and fifties.

IV. PSYCHOSEXUAL DEVELOPMENT IN LATER ADULTHOOD
   A. Developmental tasks in later adulthood include changing sexuality and loss of partner.
   B. The sexuality of older Americans tends to be invisible.
      1. Declines in sexuality among aging men and women is more cultural than biological.
      2. In American culture, sexuality and romance are more likely to be associated with the young.
      3. Sexuality is frequently linked to childbearing.
      4. Older men are physiologically less responsive than they used to be, leading to greater difficulty in attaining or maintaining an erection.
      5. Women are confronted with greater social constraints including an unfavorable gender ratio, a greater likelihood of widowhood, norms against marrying younger men, and a double standard of aging.
      6. Health and the availability of a partner are the primary determinants of an individual's sexual activity.

V. SEXUAL BEHAVIOR
   A. **Autoeroticism** includes sexual activities that involve only the self—these activities include:
      1. sexual fantasies, which are the most universal of all sexual behaviors and serve numerous psychic functions,
      2. erotic dreams, which are overtly sexual dreams and are experienced by most men and many women, and
      3. **masturbation**, which is the manual stimulation of one's own genitals.
         a. Masturbation is an important means of learning about our bodies.
         b. Whites have more liberal attitudes about masturbation than African-Americans and Latinos.
   B. Interpersonal sexuality includes a variety of sexual activities:
      1. Touching—**pleasuring** is nongenital touching and caressing permitting each partner to discover what the other likes or dislikes.
      2. Kissing is probably the most acceptable of all premarital sexual activities: It is associated with affection and attraction as well as jealousy.
      3. Oral-genital sex—**cunnilingus** is the erotic stimulation of a woman's vulva by her partner's mouth and tongue; **fellatio** is the oral stimulation of a man's penis by his partner's sucking and licking.
      4. **Sexual intercourse** or **coitus** is the insertion of the penis into the vagina and subsequent stimulation—it is associated with a complex set of interactions, meanings, and motivations.
      5. **Anal eroticism** relates to sexual activities involving the anus—**anal intercourse** is the male's insertion of his erect penis into his partner's anus.

Chapter 7

VI. **Sexual enhancement** is improving the quality of a sexual relationship, especially by providing accurate information about sexuality, developing communication skills, fostering positive attitudes, and increasing self-awareness.
   A. Enhancing our sexual relationships requires:
      1. accurate information about sexuality,
      2. an orientation toward sex based on a pleasure rather than on a performance,
      3. being in a relationship which allows a person's sexuality to flourish,
      4. an ability to communicate verbally and non-verbally about sex,
      5. being equally assertive and sensitive about sexual needs, and
      6. accepting, understanding, and appreciating differences between partners.
   B. Each person has unique conditions for good sex. These may include, but are not limited to:
      1. feeling intimate with your partner,
      2. feeling sexually capable,
      3. feeling trust,
      4. feeling aroused,
      5. feeling physically and mentally alert, and
      6. feeling positive about the environment and situation.
   C. Sexual arousal is the physiological responses, fantasies, and desires associated with sexual anticipation and sexual activity.
      1. We have different levels of arousal which are not necessarily associated with particular types of sexual activities.
      2. Intensifying erotic pleasure focuses on intensifying arousal by having one's conditions for good sex met and focusing on the sensations one is experiencing.

VII. SEXUAL RELATIONSHIPS
   A. Nonmarital sexuality
      1. **Nonmarital sex** is sexual activity, especially sexual intercourse, that takes place outside marriage.
      2. **Premarital sex**, as defined in the text, refers to younger, never-married adults under the age of 30.
      3. Over the last several decades, premarital and nonmarital sex have gained increased acceptance.
      4. **Extramarital sex** is sexual interaction that takes place outside the marital relationship: These interactions continue to receive consistent disapproval.
      5. The decision to engage in premarital sex is influenced by individual, relationship, and environmental factors.
      6. In new relationships, when sexual activities are being initiated, communication tends to be indirect and ambiguous: Traditionally, the male will initiate sexual intimacy.
      7. When beginning a sexual involvement, we need to practice safe sex and discuss birth control: Responsibility for contraception and safe sex generally requires verbal communication.

8. In comparison to marital relationships, men and women in cohabiting relationships, have sexual intercourse more frequently, are more egalitarian in initiating sexual activities, and are more likely to be involved in sexual activities outside of the relationship.
B. Gay men, because of their socialization as males, are likely to initiate sexual activity earlier in the relationship than are lesbians.
1. In both gay and lesbian relationships, the more emotionally expressive partner is likely to initiate sexual interaction.
2. Sexual exclusivity is negotiable in gay and lesbian cultures.
C. As a culture, we feel ambivalent about marital sex.
1. Sexual intercourse tends to diminish in frequency the longer a couple is married; however, most couples do not find this to be a problem if their overall relationship is good.
2. Sex within marriage differs from nonmarital sex in three ways:
a. the expectation of monogamy,
b. the socially sanctioned setting for reproduction, and
c. the context for sex in a day-to-day world—that is, sexual intercourse must be arranged around working and child care schedules.
D. Extramarital relationships violate our culture's fundamental assumption that marriages are monogamous.
1. Personal characteristics and quality of marriage appear to be the most important factors associated with extramarital relationships.
2. In general, the lower the marital satisfaction and the lower the frequency and quality of marital intercourse, the greater the likelihood of extramarital relationships.
3. Extramarital relationships exist in several forms: sexual but not emotional; sexual and emotional; and emotional but not sexual.
4. If the extramarital relationship is discovered, a marital crisis ensues.
5. Extramarital sex tends to be sporadic.
6. Most instances of extramarital relationships tend to be more sexual than emotional, although those that are more emotional create a complex system of relationships among three or four individuals.
E. **Open marriage** is a marriage in which the partners agree to allow each other to have openly acknowledged and independent sexual relationships with others: There does not appear to be a significant difference in marital stability related to whether couples are sexually open or monogamous.

VIII. SEXUAL PROBLEMS AND DYSFUNCTIONS
A. **Sexual dysfunctions** are persistent sexual problems that cause distress to the individual or the partner—some dysfunctions are physical in origin; most, however, are psychological.
1. Common dysfunctions for women include the inability to attain orgasm, arousal difficulties and dyspareunia (painful intercourse).

2. Among men, the most common dysfunctions are the inability to achieve or maintain an erection, premature ejaculation, and delayed orgasm.
   B. Origins of sexual problems can be physical or psychological.
      1. Sexual dysfunction may be physical or hormonal and may be related to various illnesses, medications, and alcohol or alcoholism.
      2. The most dominant psychological causes of sexual dysfunction are performance anxiety and conflicts within the self.
   C. Resolving sexual problems may be as simple as communicating feelings and thoughts with one's partner or friends; in other cases, seeking professional assistance may be more effective.

IX. BIRTH CONTROL
   A. **Contraception** is the prevention of pregnancy using any number of devices, techniques, or drugs.
      1. Over a year's time, a woman in a sexual relationship in which contraception is not used has an 85 to 90 percent chance of becoming pregnant.
      2. Consistent use is the key to contraceptive effectiveness: The most consistent users of contraception explicitly communicate about it.
      3. The issues in contraceptive responsibility include:
         a. acknowledging sexuality,
         b. planning contraception,
         c. obtaining contraception,
         d. continuing contraception, and
         e. dealing with the lack of spontaneity.
   B. **Abortion** is the termination of a pregnancy.
      1. The constitutional right to abortion was established by the **Roe v. Wade** decision in 1973.
      2. The characteristics of women having abortions are summarized as follows:
         a. 69 percent are white, 31 percent are nonwhite,
         b. one out of three women having an abortion is poor,
         c. 12 percent of abortions are obtained by minors, and
         d. 80 percent of adult women having abortions are separated, divorced, or never married.
      3. Women tend to have multiple reasons for having abortions.
         a. Abortion is not taken lightly: The decision is complex.
         b. A woman's developmental stage is important as well as her relationships with people and educational and economic circumstances.
      4. Men often feel powerless and forgotten in the abortion decision.
      5. It is not uncommon for couples to split up after an abortion.

X. SEXUALLY TRANSMITTED DISEASES AND HIV/AIDS
   A. **Sexually transmitted diseases** (STDs) are diseases spread through sexual contact, such as sexual intercourse, oral sex, or anal sex.

B. In the U.S., the principal STDs are chlamydia, gonorrhea, genital warts, genital herpes, syphilis, hepatitis, and HIV/AIDS.
C. HIV/AIDS
1. The **human immunodeficiency virus** (HIV) is the virus that causes AIDS.
2. **Acquired immune deficiency syndrome** (AIDS) is so named because it is a disease which people acquire that is related to the body's lack of immunity; it is a syndrome because the symptoms occur together as a group.
3. While there is no vaccine to prevent or cure HIV, there is a large body of knowledge about the virus and how to prevent its spread.
4. HIV is transported in the blood, semen, and vaginal secretions of infected persons.
5. HIV is transmitted through the exchange of blood, through sexual contact involving semen or vaginal secretions, and prenatally, from an infected woman to her fetus.
6. Heterosexuals, bisexuals, and gay men and lesbians are susceptible to the sexual transmission of HIV.
7. There is a definable progression of HIV infection and a range of illnesses associated with AIDS.
8. The presence of HIV can be detected through antibody testing. HIV antibodies develop between one and six months after infection.
9. All those with HIV are carriers regardless of whether or not they have AIDS symptoms.
D. Abstinence is the best protection from STD's and HIV: Sexually active people need to talk about STD's in an open, nonjudgemental way and use condoms.

XI. Sexual responsibility is important because we have so many sexual choices.
A. Sexual responsibility includes:
1. disclosure of intentions,
2. freely and mutually agreed-upon sexual activities,
3. use of mutually agreed-upon contraception in sexual intercourse if pregnancy is not intended,
4. use of safe-sex practices,
5. disclosure of infection from or exposure to STDs, and
6. acceptance of the consequences of sexual behavior.
B. Responsibility is facilitated when sex takes place within the context of an ongoing relationship.

XII. FEATURES AND READINGS
A. *Understanding Yourself: Your Sexual Scripts* explores the questions of who-what-when-where-why in understanding one's own sexuality.
B. *You and Your Well-Being: Choices in Unwanted Pregnancy* discusses unwanted pregnancy and the questions and issues a woman considering an abortion must deal with.
C. The *Perspective:The Abortion Debate* summarizes the "Pro-Life" and the "Pro-Choice" arguments.

Chapter 7

TEST YOUR COMPREHENSION

In the chart below, discuss the developmental tasks and the corresponding sexuality-related issues for early, middle, and later adulthood.

Chart 7.1

PSYCHOSEXUAL ADULT DEVELOPMENT		
LIFE CYCLE STAGE	DEVELOPMENTAL TASKS	SEXUALITY-RELATED ISSUES
Early Adulthood		
Middle Adulthood		
Later Adulthood		

Understanding Sexuality

The chart below illustrates the various aspects of sexuality in relationships. Define the basic characteristics of each kind of sexuality in the column under the appropriate heading.

Chart 7.2

ASPECTS OF SEXUALITY AND RELATIONSHIPS			
SEXUALITY TYPE	DEFINITION	FUNCTION	LIMITATIONS
Premarital Sexuality			
Nonmarital Sexuality			
Marital Sexuality			
Extramarital Sexuality			

113

Chapter 7

SELF QUIZZES

How well do you know this material? Test your understanding of the reading assignment by answering the following sample questions.

Part I - Multiple Choice: Choose the most correct response.

___b___ 1. Which of the following is true of parents as a source of sexual learning?
   a. Parents are usually a great deal of help for adolescents trying to understand their emerging sexuality.
   b. Parents teach their children the hidden nature of sexuality.
   c. Parents play a positive role in the development of their children's sexuality because they themselves are sexual beings.
   d. Parents are well-intentioned and effective as guides.
   e. Parents are sympathetic to sexual pressures of teens.

___d___ 2. According to the text, traditional male sexual scripts focus more on _____ than _____.
   a. feelings, sex
   b. pleasure, reproduction
   c. reproduction, pleasure
   d. sex, feelings
   e. none of the above

___e___ 3. Which is an example of a male sexual script?
   a. Men should not express certain feelings which are inappropriate.
   b. The man is in charge when it comes to sexual intercourse.
   c. A real man always wants sex and is ready for it.
   d. Sex equals orgasm; everything else is a preliminary.
   e. All of the above are examples of male sexual scripts.

___c___ 4. Contemporary sexual scripts include all of the following except
   a. sexual expression is positive.
   b. sexual activities may be initiated by either partner.
   c. women are sexual nurturers.
   d. sexual activities are a mutual exchange of erotic pleasure.
   e. All of the above are ideas from contemporary sexual scripts.

___d___ 5. The final stage of acquiring a lesbian or gay identity
   a. is marked by fear and suspicion.
   b. begins with a person's first lesbian or gay love affair.
   c. involves labeling feelings of attraction as homoerotic.
   d. involves the person's self-definition as lesbian or gay.
   e. all of the above

Understanding Sexuality

__C__ 6. All of the following are true regarding psychosexual development except
  a. men and women view aging differently.
  b. one of the greatest determinants of an aged individual's sexual activity is health.
  c. developing a sexual philosophy is a task of middle adulthood.
  d. among American women, sexual responsiveness, once reached, is usually maintained at more or less the same level into the sixties.
  e. All of the above are true.

__e__ 7. People having strong, irrational fears about gay men and lesbians
  a. often have had little contact with them.
  b. often have a deeply rooted insecurity concerning their own sexuality and gender identity.
  c. often have a strong fundamentalist religious orientation.
  d. often are ignorant about gay men and lesbians.
  e. all of the above

__e__ 8. Good sex is enhanced by
  a. being aware of one's own sexual needs.
  b. verbal and nonverbal communication.
  c. appreciating differences between partners.
  d. an orientation toward sex based on pleasure rather than performance.
  e. all of the above

__c__ 9. Which of the following is not true regarding sexual fantasies?
  a. Fantasies help direct and define our erotic goals.
  b. Erotic fantasizing is probably the most universal of all sexual behaviors.
  c. Fantasies are representative of what a person does sexually in real life.
  d. Fantasies provide a form of rehearsal, allowing us to practice in our minds.
  e. Fantasy offers a safe outlet for sexual curiosity.

__d__ 10. Which of the following is not true about masturbation?
  a. It is one way that people learn about their bodies and what pleases them.
  b. It is almost a universal experience for men.
  c. It serves the function of release of sexual tension, especially when a partner is not available.
  d. It is usually viewed as bad and harmful by those who practice it.
  e. Whites are more accepting of masturbation than African-Americans and Latinos.

__c__ 11. Which of the following is not true of "pleasuring"?
  a. It may involve the pleasure of touch, smell, sound, taste, and sight.
  b. It is a way by which couples can get to know each other and each other's bodies.
  c. It always leads to orgasm.
  d. It is nongenital touching and caressing.
  e. It has been largely ignored as a sexual behavior.

Chapter 7

____ 12. Which of the following is not true about marital sexuality?
   a. There is currently considerable research being done on marital sexual relationships.
   b. Sex is only one bond among many for married people.
   c. Sex within marriage is expected to be monogamous.
   d. Sex within marriage takes on a procreative meaning.
   e. Marital sex tends to diminish in frequency the longer a couple is married.

____ 13. People are more likely to become involved in affairs because
   a. the marital satisfaction is low.
   b. the frequency and satisfaction of marital sexual intercourse is low.
   c. they feel something vital is missing in their relationship with their spouse.
   d. affairs act as compensation or a substitute for the deficiencies in a marriage.
   e. all of the above

____ 14. Which of the following is likely to not be a sexual problem among men?
   a. dyspareunia
   b. erectile dysfunctions
   c. premature ejaculation
   d. delayed orgasm
   e. inhibited sexual desire

____ 15. People take chances by having intercourse without using contraceptives because
   a. they underestimate how easy it is to get pregnant.
   b. our society obstructs rather than encourages contraception.
   c. using contraceptives acknowledges that one is sexual.
   d. they are in a casual dating relationships.
   e. all of the above

____ 16. Which of the following is not true regarding sexually transmitted diseases?
   a. You cannot tell by a person's looks whether he or she is carrying a sexually transmitted disease.
   b. Untreated sexually transmitted diseases may lead to pelvic inflammatory disease and even sterility for women.
   c. Sexually transmitted diseases happen even to people who are "nice."
   d. Because of penicillin and other drugs, sexually transmitted diseases are on the decline in the United States.
   e. College students are as vulnerable as any other group to catch sexually transmitted diseases.

____ 17. Which of the following is not true regarding AIDS?
   a. AIDS is caused by the human immunodeficiency virus (HIV) that attacks the body's immune system.
   b. Sexually transmitted cases of AIDS among heterosexuals are increasing at a greater rate than among gay men.
   c. If a person carries the virus, even though no symptoms appear, he or she can still infect other people.
   d. The human immunodeficiency virus actually kills people directly.
   e. AIDS is transmitted only in certain clearly defined circumstances.

Understanding Sexuality

Part II - True/False

__F__ 1. Peer groups are a reliable source of sexual information for most teens in our culture.

__T__ 2. Women are taught that sex is both good and bad.

__F__ 3. Male sexual scripts merge sex and love.

__F__ 4. Homoeroticism occurs only as a result of gay or lesbian activity.

__T__ 5. Heterosexuals view bisexuals as really homosexual.

__F__ 6. Menopause is a sudden event.

__T__ 7. Masturbation can be a form of emotional protection.

__T__ 8. Among couples, cross-culturally, kissing a person other than the partner evokes jealousy.

__T__ 9. Good sex does not necessarily include orgasm or intercourse.

__T__ 10. According to research, men perceive slightly more pressure or obligation to engage in intercourse than women.

__T__ 11. Marriage has lost some of its power as the only legitimate setting for sexual intercourse.

__T__ 12. Marital sex tends to diminish in frequency the longer a couple is married.

__T__ 13. A person who had premarital sex is more likely to have extramarital sex.

__F__ 14. In a recent study by Rubin and Adams, sexually open marriages were found to be significantly less stable than monogamous marriages.

__F__ 15. There is about a fifty percent chance that a fertile woman will get pregnant within a year if she and her partner use no contraceptives.

__F__ 16. With the advent of women's liberation and better sex education in the schools, most teenagers are well-informed on contraceptives and use them conscientiously.

__T__ 17. Related to abortion, men often feel forgotten and powerless.

__T__ 18. Americans are in the middle of the worst STD epidemic of our history.

117

Chapter 7

DISCUSS BRIEFLY

1. In what ways is our society homophobic? What are the effects of homophobia on gays and lesbians?

2. How does homophobia affect our relationships and friendships with individuals of our same sex?

3. In what ways do gay and lesbian relationships differ from traditional heterosexual ones? In what ways are they the same?

4. What are some of the sexual messages men have received and how are they dysfunctional?

5. What are some of the sexual messages women have received and how are they dysfunctional?

SELF-DISCOVERY

What kind of sex education did you receive? Who was involved in this education? Did you feel that this was adequate? Would you educate your own child or children (if you had any) differently?

Chapter 7

Do you think it is possible for a loving relationship to be "open" or non-monogamous under certain circumstances? If so, under what circumstances and with what understandings? If not, why? Why is it difficult for such "open" relationships to work?

KEY TO SELF QUIZZES

Multiple Choice **True/False**

1. b	10. d	1. F	10. T
2. d	11. c	2. T	11. T
3. e	12. a	3. F	12. T
4. c	13. e	4. F	13. T
5. b	14. a	5. T	14. F
6. c	15. e	6. F	15. F
7. e	16. d	7. T	16. F
8. e	17. d	8. T	17. T
9. c		9. T	18. T

SUGGESTED READINGS

For related readings see page 253 in your text

CHAPTER 8

Pregnancy and Childbirth

MAIN FOCUS

Chapter Eight examines pregnancy tests, pregnancy and childbirth as biological, emotional, and social processes, sexuality during pregnancy, complications of pregnancy, diagnosing fetal abnormalities, and pregnancy loss. It also discusses giving birth, labor and delivery, choices in childbirth, the medicalization of childbirth, circumcision, breastfeeding, and becoming a parent.

GOALS OF THIS CHAPTER

To demonstrate mastery of this chapter, you should be able to:

1. Discuss the biological, emotional, social, and psychological aspects of pregnancy.

2. Discuss guidelines and options related to sexuality during pregnancy.

3. Describe the various complications of pregnancy and dangers to the fetus.

4. Explain the procedures for diagnosing abnormalities of the fetus.

5. Understand and describe pregnancy loss and infant mortality, and the subsequent grief and healing processes.

6. Describe the stages of normal labor and delivery and discuss birth by cesarean section.

7. Explain how childbirth is a social experience and be familiar with the various choices in childbirth, including natural childbirth, birth place options, and the role of the midwife.

Chapter 8

8. Discuss the impact of the medicalization of childbirth on prospective parents.

9. Explain the debate over circumcision.

10. Describe the physical, psychological, and emotional aspects of breastfeeding.

11. Explain the postpartum period and the transition to parenthood.

12. Describe the social and emotional effects of the childbirth process on the father.

13. Be familiar with reproductive technology and the moral issues that are emerging around it.

KEY TERMS AND IDEAS

The following terms, ideas, and concepts are listed in the order in which they appear in Chapter Eight. Be sure that you understand and can define each of the following:

Hegar's sign	chorionic villus samplings (CVS)	neonate
teratogens	alpha-feto protein (AFP) screening	Apgar score
ectopic pregnancy	sudden infant death syndrome (SIDS)	involution
toxemia	parturition	lochia
preeclampsia	Braxton Hicks contractions	episiotomy
eclampsia	effacement	cesarean section (C-section)
low birth weight (LBW)	dilation	prepared childbirth
ultrasonography	transition	circumcision
sonogram	vernix	lactation
amniocentesis	placenta	colostrum
		postpartum period

CHAPTER EIGHT OUTLINE

I. INTRODUCTION TO THE CHAPTER
   A. The birth of a wanted child is considered by many parents as a happy occasion.
   B. Today, pain and controversy surround many aspects of pregnancy and childbirth.
   C. Chapter Eight views pregnancy and childbirth from biological, social, and psychological perspectives.

II. BEING PREGNANT
   A. Pregnancy tests detect the presence of human chorionic gonadotropin (HCG) by use of an agglutination test; a little later, **Hegar's sign** is visible.
   B. A woman's feelings during pregnancy will vary dramatically depending on how she feels about pregnancy and motherhood, whether the pregnancy was planned, and how secure her home situation is.
      1. A woman's first pregnancy is especially important because it has traditionally symbolized the transition to maturity.

2. The pregnant woman's relationships to her partner and to her mother are likely to change during pregnancy.
3. Over the course of the pregnancy the woman experiences a great many changes.
   a. The first trimester may involve fatigue, nausea, sore breasts, and fearfulness or anxiety.
   b. The second trimester brings respite from early symptoms and the feeling of fetal movement.
   c. The third trimester may be the time of greatest difficulties in daily living, with possible edema, physical limitations, worries about childbirth, depression, feeling physically awkward and sexually unattractive. Many experience excitement and anticipation accompanied by bursts of energy.
4. Both partners have important developmental tasks which will need to be resolved.

C. A woman's sexual feelings and actions fluctuate during pregnancy: Men may feel confusion or conflicts about sexual activity during pregnancy.
1. Although there are no "rules" governing sexual behavior during pregnancy, basic precautions should be observed.
2. Guidelines for expressing sexual feelings during pregnancy include trying new positions, considering alternatives to intercourse such as masturbation and cunnilingus, and remember there are no rules about sexuality and pregnancy.

D. Usually pregnancy proceeds without major complications; however, there are several potential complications of pregnancy.
1. Harmful substances may reach the embryo or fetus through the placenta.
   a. **Teratogens** (substances that cause defects in developing embryos or fetuses) are directly traceable in only 2 or 3 percent of the cases of birth defects; they are thought to be linked to 25 to 30 percent of such cases.
   b. The ingestion of alcohol during pregnancy can lead to fetal alcohol syndrome (FAS) or, in lesser amounts, to fetal alcohol effect (FAE).
   c. Pregnant women who regularly use opiates (heroine, morphine, codeine, and opium) are likely to have infants who are addicted at birth: Many drug-exposed infants have been subjected to alcohol exposure as well.
   d. Cigarette smoking affects the unborn child and has been implicated in sudden infant death syndrome, respiratory disorders in children, and various adverse pregnancy outcomes.
   e. Caffeine, prescription drugs, vitamins, chemicals, and environmental pollutants are also potentially threatening to the unborn child.
2. Infectious diseases such as German measles (rubella), Group B streptococcus, and sexually transmitted diseases can damage the fetus or infant.
3. Over 6,000 children under age 5 have been diagnosed with AIDS. The drug AZT however, significantly reduces the transmission of HIV to the fetus.
4. Herpes simplex may cause brain damage or death for infants.
5. Complications specific to pregnancy include **ectopic pregnancy, toxemia, preeclampsia** and **eclampsia**, and prematurity or **low birth weight (LBW)**.

Chapter 8

      6. Premature delivery (LBW) is one of the greatest problems confronting obstetrics today.
         a. About half the cases are related to teenage pregnancy, smoking, poor nutrition and poor health in the mother.
         b. Prenatal care is extremely important as a means of preventing prematurity.
   E. Technology and tests used in diagnosing abnormalities of the fetus include **ultrasonography, sonograms, amniocentesis, chorionic villus sampling (CVS), and alpha-feto protein (AFP) screening**.
   F. The loss of a child through miscarriage, stillbirth, or death during early infancy is a devastating experience that has been largely ignored in our society.
      1. Spontaneous abortion (miscarriage) is a powerful natural selective force toward bringing healthy babies into the world: 60% of all miscarriages are due to chromosomal abnormalities.
      2. Among developed nations, our country ranks twenty-first for low infant mortality. Twenty countries have lower infant mortality rates than the United States.
         a. Most of the babies who die less than one year old are victims of the poverty that often results from racial or ethnic discrimination.
         b. The United States is far behind many other countries in terms of providing health care for children and pregnant women.
         c. The U.S. infant mortality rate is also related to congenital birth defects, infectious diseases, accidents, and other causes.
         d. Sometimes the causes of death are not apparent; **sudden infant death syndrome (SIDS)** is a particularly perplexing phenomenon wherein an apparently healthy infant dies suddenly while sleeping.
   G. The death of the fetus can represent the death of a dream and of a hope for the future.
      1. The loss must be acknowledged and experienced before psychological healing can take place.
      2. Women who lose a pregnancy generally experience similar stages in their grieving process: Initially shock and numbness, then sadness, guilt, and anger.
      3. Healing takes time—months, a year, perhaps more.
      4. Support groups and counseling are often helpful, especially if healing does not seem to be progressing.

III. GIVING BIRTH
   A. Throughout pregnancy, numerous physiological changes occur to prepare the woman's body for childbirth or **parturition**. In the third trimester **Braxton Hicks contractions** exercise the uterus preparing it for labor.

B. During labor these contractions begin the **effacement** (thinning) and **dilation** (opening up) of the cervix.
   1. The first of the three stages of labor is usually the longest, and may include expulsion of the birth plug, loss of amniotic fluid (which comes from the ruptured amniotic membrane), and increasing uterine contractions, lasting until **transition**. At the end of the first stage, the baby's head is poised to enter the birth canal.
   2. The second stage begins with the baby's head moving down the birth canal and ends with the birth of the baby, covered with **vernix**, and still attached to the mother by the umbilical cord.
   3. The expelling of the **placenta** (afterbirth) is the third stage.
   4. Immediately following the birth, the attendants assess the physical condition of the **neonate** and assign an **Apgar score**.
   5. For a few days following labor, the mother will probably feel strong contractions as the uterus begins to return to its prebirth size and shape.
      a. This process, called **involution**, takes about six weeks.
      b. She will also have a bloody discharge (**lochia**) which continues for several weeks.
C. Traditional hospital birth has been criticized as impersonal, and designed for the convenience of the obstetrician rather than for the comfort of the woman.
   1. Some hospitals are responding to this criticism by focusing on family-centered births.
   2. Some form of anesthetic is administered during most hospital births.
   3. During delivery, the mother will probably be given an **episiotomy**.
D. **Cesarean section (C-section)** is the removal of the fetus by an incision in the mother's abdominal and uterine walls.
   1. There has been an alarming increase in the number of C-sections performed in the U.S., and the National Institutes of Health have taken steps to try to reduce the number.
   2. Since the C-section rate for middle- and upper-income women is seven times the rate for lower-income women, it is assumed that cesareans are often performed for reasons other than medical risk.
E. **Prepared childbirth** (or natural childbirth) involves the woman using few or no drugs, reducing fear and tension, and using various techniques to decrease pain: Clinical studies consistently show better birth outcomes for mothers who have had prepared childbirth classes.
F. Alternatives to traditional hospital births include birthing rooms and centers, home birth, and midwifery.
G. The medicalization of childbirth leads us to assume that childbirth is an inherently dangerous process when, in fact, only 5 to 10 percent of births actually require medical procedures: Society's increasing dependence on technology may hamper women's ability to view childbirth as a natural process for which they are naturally equipped.
H. In spite of the fact that it is not medically necessary, almost sixty percent of newborn American boys are still experiencing routine **circumcision** in comparison to less than one percent of the newborn boys in other Western countries.

Chapter 8

   I. The American Academy of Pediatricians endorses total breastfeeding for a baby's first six months.
      1. **Lactation** begins about three days after birth; prior to this a liquid called **colostrum** which is highly nutritious is secreted by the nipples.
      2. Breastfeeding, increasingly popular, has important advantages over bottle feeding for both infant and mother.
         a. A healthy mother, consuming a good diet, offers the best nutrition for the baby.
         b. Nursing provides a sense of emotional well-being for both mother and child.
         c. Bottle feeding an infant makes it possible for the father to share in the nurturing process.

IV. BECOMING A PARENT
   A. The irrevocable nature of parenthood may make the first-time parent doubtful and apprehensive, especially during pregnancy.
   B. No amount of preparation can fully prepare a couple for life with a new baby.
   C. The **postpartum period** is a time of physical stabilization and emotional adjustment during which some women may face "postpartum blues" and sleeplessness.
   D. Biologically, postpartum women experience an abrupt fall in certain hormone levels, may also be dehydrated, and suffer blood loss.
   E. Psychologically, a woman may have to deal with her ability to mother, ambivalent feelings toward her own mother, and communication problems with her infant or partner.
   F. Social factors such as finances and the emotional stress of other family members are also significant.
   G. Postpartum counseling prior to discharge from the hospital can help couples gain perspectives and evaluate their resources.
   H. Men appear to experience postpartum depression as well when they do not feel prepared for parenting and financial responsibilities.
      1. Some men feel overwhelmed by changes in the marital relationship.
      2. Fatherhood is a major transition for them, but their feelings are overlooked.
   I. The transition to parenthood can be made easier if new parents expect a certain amount of tiredness and stress, ascertain sources for support, keep communication open, and plan time to be together.
   J. For many women and men, the arrival of a child is one of life's most important events which is accompanied by a deep sense of accomplishment.

V. FEATURES AND READINGS
   A. *Other Places... Other Times* describes the different activities in different cultures regarding the practice of couvade, imitating both pregnancy and childbirth.
   B. *You and Your Well-Being: Planning for Pregnancy: Preconception Care* discusses the recommendations of the U.S. Public Health Service that women prepare for pregnancy by considering their general health, age, nutrition, substance consumption, and family health history and then take appropriate actions before conceiving.
   C. In the *Ethics of Reproductive Technology* the reader is asked to evaluate and assess the possible benefits of each technique against the costs and risks involved.

TEST YOUR COMPREHENSION

Below is a chart illustrating some of the ethical issues surrounding reproductive technology. Describe the ethical and/or social ramifications of each issue, using readings, lecture and personal research.

Chart 8

ETHICAL & SOCIAL ISSUES SURROUNDING PREGNANCY AND CHILDBIRTH	
Issue	**Ethical and/or Social Implications**
Fetal Diagnosis	
Cesarean Section	
Intrauterine Insemination and In Vitro Fertilization	
Older Motherhood	
Surrogate Motherhood	
Abortion	
Human Cloning	

Chapter 8

SELF QUIZZES

How well do you know this material? Test your understanding of the reading assignment by answering the following sample questions.

Part I - Multiple Choice: Choose the most correct response.

___a___ 1. The presence of human chorionic gonadotropin (HCG) is associated with
   a. agglutination tests.
   b. Hegar's sign.
   c. lochia.
   d. preeclampsia.
   e. infertility.

___e___ 2. Which of the following is a developmental task for both the expectant mother and the expectant father?
   a. resolution of dependency issues
   b. evaluation of financial issues
   c. attachment to the fetus
   d. acceptance and resolution of the relationship with the same sex parent
   e. all of the above

___c___ 3. One of the most important factors in having a complication-free pregnancy is
   a. refraining from sexual intercourse.
   b. taking vitamins in large amounts.
   c. good nutrition.
   d. elevating teratogens.
   e. all of the above

___b___ 4. Which of the following is least likely to be harmful to the developing fetus through the placenta?
   a. chronic ingestion of alcohol
   b. caffeine from a cup of coffee
   c. cigarette smoking
   d. accutane
   e. opiates

___c___ 5. All but which one of the following is true regarding infectious diseases and pregnancy?
   a. Gonorrhea may expose the baby to blindness.
   b. Herpes simplex may cause brain damage and infant death.
   c. All infants born HIV positive remain HIV positive and are at high risk of dying from AIDS.
   d. In recent years the rates of congenital syphilis have increased dramatically.
   e. All of the above are true.

128

Pregnancy and Childbirth

___e___ 6. Which of the following complications of pregnancy is characterized by high blood pressure?
   a. ectopic pregnancy
   b. preeclampsia
   c. eclampsia
   d. toxemia
   e. all but a

___e___ 7. Prematurity or low birth weight is likely to correlate with
   a. teen pregnancy.
   b. smoking.
   c. poor nutrition.
   d. poor health of the mother.
   e. all of the above

___a___ 8. Which of the following procedures has the lowest risk to the fetus?
   a. ultrasonography
   b. amniocentesis
   c. chorionic villus biopsy
   d. surgical treatment in the womb
   e. All of the above have low rates of risk to the fetus.

___d___ 9. Which of the following is not true regarding infant mortality rates in the U.S.?
   a. Infant mortality rates are twice as high for blacks compared to whites.
   b. High infant mortality rates are often the result of social factors such as poverty and unemployment.
   c. Infant mortality rates are affected by poor nutrition and substance abuse.
   d. Infant mortality rates should improve significantly now that we are more knowledgeable about the social causes of high death rates and have more programs to prevent them.
   e. Some cities of the United States have infant death rates similar to those of underdeveloped countries.

___d___ 10. Uterine contractions which prepare the uterus for labor are called
   a. effacement.
   b. dilation.
   c. preeclampsia.
   d. Braxton Hicks.
   e. none of the above

___b___ 11. Transition, during which the baby's head enters the birth canal, occurs during
   a. the beginning of first-stage labor.
   b. the end of first-stage labor.
   c. the beginning of second-stage labor.
   d. the end of the second-stage labor.
   e. the third-stage labor.

Chapter 8

___c___ 12. Which of the following is true regarding cesarean sections?
   a. The woman usually knows in advance that it will be necessary.
   b. If a woman once has a cesarean section, all of her subsequent deliveries must be by C-section as well.
   c. Mothers have a significantly higher mortality rate in cesarean births.
   d. Breech babies always require a C-section.
   e. Due to increasing sophistication of birth techniques, fewer cesarean sections are done each year.

___e___ 13. Prepared childbirth involves
   a. reducing fear which causes muscles to tense.
   b. the woman psychologically separating herself from the conditioned responses of pain and anxiety.
   c. reducing the use of drugs during labor and delivery.
   d. the increased participation of many fathers in the pregnancy and birth process.
   e. all of the above

___b___ 14. Prepared childbirth
   a. is painless.
   b. means that the partners understand what is occurring at all stages of labor and delivery.
   c. means home delivery with a midwife.
   d. teaches that childbirth is only a woman's experience.
   e. may increase fear when you find out how "awful" having a baby is.

___c___ 15. Each of the following is true regarding circumcision except
   a. it is often done without anesthesia.
   b. the practice grew out of the belief that any boy who washed his penis would discover masturbation.
   c. a circumcised penis is cleaner than an intact penis.
   d. some parents have it done so the son will "look like his dad".
   e. there is no medical indication for or against circumcision.

___a___ 16. Which of the following is not true of breastfeeding?
   a. Almost all women breastfeed today.
   b. Breast milk contains antibodies provided by the mother which helps the baby retain immunity to illness.
   c. Breastfeeding has psychological benefits for both mother and child.
   d. A mother who is breastfeeding should check with her doctor before taking any drugs or medication.
   e. Breast milk is easier to digest for the premature baby.

Pregnancy and Childbirth

__ae__ 17. Postpartum depression is usually a result of
   a. irregular sleep patterns.
   b. an abrupt fall in certain hormonal levels.
   c. physiological stress accompanying labor, dehydration, blood loss, and decrease in stamina.
   d. feelings of helplessness and ambiguity about one's ability to mother.
   e. all of the above

__a__ 18. According to the text, _____ is a dramatic symbol of a man's paternity and his magical relation to the child.
   a. couvade
   b. impregnating a woman
   c. Hegar's sign
   d. psychological gestation
   e. an apgar score

__e__ 19. Which of the following may illustrate an ethical issue of reproductive technology?
   a. in vitro fertilization
   b. older motherhood
   c. surrogate motherhood
   d. amniocentesis
   e. all of the above

Part II - True/False

__F__ 1. Home pregnancy tests offer an over-the-counter version of alpha-feto protein screening.

__T__ 2. Pregnant women may change their relationships with their own mothers dramatically, sometimes being more independent, other time reconciling estranged relationships.

__T__ 3. Orgasms can induce uterine contractions.

__T__ 4. Pregnant women who regularly use opiates are likely to have infants who are addicted at birth.

__T__ 5. Some infants born HIV positive reconvert to negative status.

__F__ 6. Almost all pregnant women who have had Herpes Simplex II will have to deliver their babies via cesarean section.

__T__ 7. Many low birth weight babies will experience disabilities.

__T__ 8. More than half of all miscarriages are due to defects in the fetus and as many as one third of all implanting embryos miscarry.

Chapter 8

___F___ 9. The United States is a leader among developed nations in terms of providing health care for children and pregnant women.

___F___ 10. The second stage of labor is often the longest and the most uncomfortable.

___F___ 11. An episiotomy is generally performed to help an infant experiencing distress.

___T___ 12. Cesarean sections are often performed for reasons other than medical risk.

___T___ 13. Women who have prepared for childbirth have been found to: have reduced pain perception and increased endurance, cooperate more with the physician, and describe their childbirth experiences as more satisfying.

___F___ 14. According to the text, home births have been found to be significantly more risky than births at hospitals.

___T___ 15. The idea that the pain of childbirth is to be avoided at all costs is a relatively new one.

___T___ 16. The American Academy of Pediatrics endorses total breastfeeding for a baby's first six months.

___F___ 17. There are strong medical reasons for circumcising male children.

___T___ 18. Men, as well as women, appear to experience postpartum blues.

___F___ 19. Pregnancy and childbirth generally have minimal psychological effects on the father.

DISCUSS BRIEFLY

1. What factors do a woman and man have control over in terms of producing a healthy child?

2. What are some of the ways that couples are currently changing the roles concerned with birthing and parenting? How have these choices particularly affected men?

3. What are some of the advantages of choosing to breastfeed a baby? Some of the disadvantages? How do you feel about it?

Chapter 8

SELF-DISCOVERY

1. As an alternative to the impersonal, routine hospital birth, would you consider a birth center? home birth? midwifery? Why or why not?

2. Would you consider surrogate motherhood if it was your only means of conceiving a child? Why or why not?

3. If you (or your partner) had amniocentesis and it was determined that the fetus was suffering severe mental retardation, would you consider abortion? Why or why not?

MINI-ASSIGNMENT I

It has been suggested that our society is very "pro-natal" and that we are constantly being bombarded with media messages that support having babies, when in fact many people would be better off or really prefer to be "childfree." Can you give some examples of messages our culture gives us about how wonderful it is to have babies?

Collect illustrations from magazines which glorify motherhood and pregnancy. Are there contrasting illustrations which show the problems of caring for infants? How realistic do you think these illustrations are?

Chapter 8

MINI-ASSIGNMENT II

Ask your friends, relatives, and acquaintances about the effects of the birth of their first baby on the new parents' marriage. Share the information that you have gathered with your classmates. What information have they gathered? How many of the effects are positive? How many are negative?

KEY TO SELF QUIZZES

Multiple Choice		True/False	
1. a	10. d	1. F	10. F
2. e	11. b	2. T	11. F
3. c	12. c	3. T	12. T
4. b	13. e	4. T	13. T
5. c	14. b	5. T	14. F
6. e	15. c	6. F	15. T
7. e	16. a	7. T	16. T
8. a	17. e	8. T	17. F
9. d	18. a	9. F	18. T
	19. e		19. F

SUGGESTED READINGS

For related readings, see page 285 in the text.

CHAPTER 9

Marriage as Process: Family Life Cycles

MAIN FOCUS

Chapter Nine examines the stages of the individual and family life cycle from beginning marriages to later-life marriages and widowhood. In addition, it examines marital satisfaction, grandparenting, and enduring marriages.

GOALS OF THIS CHAPTER

To demonstrate mastery of this chapter, you should be able to:

1. Explain how an individual may experience living as a process through a life cycle.

2. Identify the traditional stages of the family life cycle and the effects of social change in creating family life cycle variations.

3. Describe and detail the family life cycle from beginning families to aging families.

4. Discuss the traditional assumptions about marital responsibility, the changes in those assumptions, and the primary adjustment tasks which newly married couples must undertake.

5. Discuss marital satisfaction over the family life cycle.

6. Explain the functions of engagement and cohabitation.

7. Describe the empty nest and the not-so-empty nest.

8. Discuss contemporary grandparenting, including the three styles of grandparenting.

9. Discuss factors related to marital satisfaction.

Chapter 9

KEY TERMS AND IDEAS

The following terms, ideas, and concepts are listed in the order that they appear in Chapter Nine and in the outline. Be sure that you understand and can define each of the following:

    family life cycle   empty-nest syndrome
    honeymoon effect   boomerang generation
    duration-of-marriage   intermittent extended families
    identity bargaining   sandwich generation

CHAPTER NINE OUTLINE

I. INTRODUCTION TO THE CHAPTER
    A. In the process of marriage, people interact with each other, create families, and give each other companionship and love.
    B. Marriage is not static; it is always changing to meet new situations, new emotions, new commitments, and new responsibilities.

II. THE DEVELOPMENTAL PERSPECTIVE
    A. The developmental perspective sees individual and family development as interacting with each other.
        1. Erik Erikson describes the human life cycle as containing eight development stages: infancy, toddler, early childhood, school age, adolescence, young adulthood, adulthood, maturity.
        2. At each stage, we have an important developmental task to accomplish: The way we deal with these stages is strongly influenced by our families, marriages, or other intimate relationships.
    B. The concept of the **family life cycle** uses a developmental framework to explain people's behavior in families. At various stages in the family life cycle, the family has different developmental tasks to perform.
        1. Families change over time in terms of both who are members of the family and the roles they play.
        2. The key factor in developmental studies is the presence of children; the family organizes itself around its child rearing responsibilities.
    C. Duvall and Miller have described eight stages of the family life cycle: beginning families; childbearing families; families with preschool children; families with schoolchildren; families with adolescents; families as launching centers; families in the middle years; and aging families.
    D. A major limitation of the family life-cycle approach is its tendency to focus on the intact nuclear family as "the family."

1. Family life-cycle variations include: deferred marriages, cohabitation, divorce, remarriage, single parenthood, lifelong singles, gay and lesbian families, families with disabilities, and ethnic families.
2. Researchers are increasingly focusing on two common alternatives to the traditional intact family life cycle: the single-parent and the stepfamily life cycles.

III. BEGINNING MARRIAGES
   A. Americans are waiting longer to marry today than in previous generations.
   B. Increasing age at time of marriage probably results in young adults beginning marriage with more maturity, independence, work experience, and education.
   C. The time and patterns that precede marriage can often predict how happy a couple will be in marriage.
      1. The notion that marriage will change a person for the better is a dangerous myth.
      2. Background factors which appear to marital happiness include: age at marriage; length of courtship; level of education; and childhood environment.
      3. Opposites repel: We choose partners who share similar personality characteristics.
      4. Personality does affect marital processes.
      5. Research suggests that negative interactions did not significantly affect the first year of marriage due to the **honeymoon effect**.
   D. The first stage of the family life cycle may begin with engagement or cohabitation, followed by a wedding, the ceremony that represents the beginning of a marriage.
      1. Today, more couples than in the past announce that they are "planning to get married"— this lacks the formality of engagement and is less socially binding.
      2. Engagement performs several functions:
         a. It signifies a commitment to marriage.
         b. It prepares couples for marriage by requiring them to think about the realities of everyday married life.
         c. It is the beginning of kinship.
         d. It allows the prospective partners to strengthen themselves as a couple.
      3. The rise of cohabitation has led to its becoming an alternative beginning of the contemporary family life cycle.
      4. The wedding is an ancient ritual that symbolizes a couple's commitment to each other.
         a. The central meaning of the wedding is that it symbolizes a profound life transition.
         b. The man and the woman take on marital roles.
         c. For young men and women entering marriage for the first time, marriage signifies a major step into adulthood.
         d. A wedding is a highly significant rite of passage.
   E. The expectations that two people have about their own and each other's marital roles are based on gender roles and their own experience.

Chapter 9

1. Traditional marital roles have four assumptions about husband/wife responsibilities:
   a. The husband is the head of the household.
   b. The husband is responsible for supporting the family.
   c. The wife is responsible for domestic work.
   d. The wife is responsible for child rearing.
2. Although there have been significant changes over the last generation concerning gender and marital roles, many expectations have not significantly changed.

F. There are several marital tasks that newly married couples need to begin in order to build and strengthen their marriages: Failure to successfully complete these tasks may contribute to the **duration-of-marriage effect**.
   1. Duration-of-marriage effect refers to the accumulation over time of various factors that cause marital disenchantment.
   2. Primary marital adjustment tasks include:
      a. establishing marital and family roles;
      b. providing emotional support;
      c. adjusting personal habits;
      d. negotiating gender roles;
      e. making sexual adjustments;
      f. establishing family and employment priorities;
      g. developing communication skills;
      h. establishing kin relations; and
      i. participating in the larger community.
   3. **Identity bargaining** is the process by which people adjust their idealized pre-conceptions of marriage to the reality of their partner's personality and to the circumstances of their marriage.
   4. Another important task for the new couple is establishing boundaries.
      a. Newly married couples must decide how much interaction with their families is desirable and how much influence their families of orientation may have.
      b. The critical task is to form a family that is interdependent, rather than totally independent or dependent.

IV. YOUTHFUL MARRIAGES
   A. Youthful marriages represent stages II to IV in the family life cycle: childbearing families; families with preschool children; and families with school children.
   B. In general, the birth of the first child has little impact on the husband's relationship with his work.
   C. The wife's life alters radically with the birth of the first child and she may have to make considerable psychological adaptation in her transition to motherhood.
   D. Typical struggles in families with young children concern child-care responsibilities and parental roles.

E. The impact of children may cause the partners to increasingly grow apart during this period.
F. For adoptive families, the transition to parenthood may differ from that of biological families.

V. MIDDLE-AGED MARRIAGES
A. Middle-aged marriages generally represent stages V and VI in the family life cycle: families with adolescents and families as launching centers.
B. Families with adolescents, require considerable family reorganization on the part of the parents.
1. Increased family conflict may occur as adolescents begin to assert their autonomy and independence.
2. Despite the growing pains accompanying adolescence, parental bonds generally remain strong.
C. As the families become launching centers, the parental role becomes increasingly less important in daily life: Marital satisfaction usually begins to rise for the first time since the first stage of marriage.
D. Traditionally, it has been asserted that the departure of the last child from home leads to **an empty nest syndrome** among women, characterized by depression and identity crises: There is little evidence that such a syndrome is widespread.
E. The number of twenty-five year olds living with one or both parents rose from 15 percent in 1970 to 21 percent in 1990.
1. The **boomerang generation** are those who do an extra rotation through their family home after a temporary or lengthy absence.
2. Ethnic and racial tradition seem to be more important than family income in determining whether a young adult will leave home.
F. Couples in middle age tend to reexamine their aims and goals.

VI. LATER-LIFE MARRIAGES
A. Later-life marriages represent stages VII and VIII of the family cycle: A later-life marriage is one in which the children have been launched and the partners are middle-aged or older.
1. The three most important factors affecting middle-aged and older couples are health, retirement, and widowhood.
2. These men and women must often assume caretaking roles of their own aging parents or adjust to adult children who have returned home.
B. As they enter old age, men and women are better off than we have been accustomed to believe.
1. Beliefs that the elderly are neglected and isolated tend to reflect myth more than reality.
2. Poverty rate for the elderly declined during the 1980s while that of other groups substantially increased.

Chapter 9

   3. The health of the elderly also appears to be improving as they increase their longevity.
   4. More married couples are living into old age, and there are fewer widows at younger ages.
C. Some families may become **intermittent extended families** during their later-life stages.
   1. These are families that take in other relatives temporarily during a time of need.
   2. Intermittent extended families tend to be linked to ethnicity.
D. The **sandwich generation** is a relatively new phenomena in which middle-age individuals are sandwiched between the simultaneous responsibilities of raising both their dependent children and their aging parents.
E. Most elderly parents still feel themselves to be parents, but their parental role is considerably less important in their daily lives.
   1. They generally have regular contact with their adult children, usually by letters or phone calls.
   2. Parents tend to assist those percieved to be in need, especially children who are single or divorced.
   3. They often provide financial assistance or other services to their children: Adult children generally reciprocate in terms of physical energy.
   4. Despite their continued parental concern, they tend to be maritally rather than parentally oriented.
F. Grandparents are a very present aspect of contemporary American family life.
   1. Grandparenting is expanding tremendously these days, creating new roles that relatively few Americans played a few generations back.
   2. Grandparents play important emotional roles in American families, which usually involve strong bonds.
   3. Kennedy and Kennedy (1993) found that the significance of grandparents varies by family form: Grandparents seem to take on even greater importance in single-parent and step-parent families.
   4. There tends to be three distinct styles of grandparenting:
      a. Companionate relationships are marked by affection, companionship, and play.
      b. Remote grandparents are not intimately involved in their grandchildren's lives due to geographical remoteness rather than emotional remoteness.
      c. Involved grandparents are actively involved in parenting activities with their grandchildren.
   5. Stepgrandparents are often confused about their grandparenting role.
G. Retirement should be viewed as a process: Retiree's experiences are often different and multi-layered.
   1. The key to marital satisfaction in later years is continued good health.
   2. There is a growing trend toward early retirement among the financially secure.
   3. The role of husband becomes more important as men focus on leisure activities with their wives: Men participate in more household activities than they did when they were working.

Marriage as a Process: Family Life Cycle

    4. Women who retire may not find themselves facing the same identity issues as men because women have often had important family roles which continue.
    5. In retirement, the marital relationship continues along the same track as prior to retirement.
    6. The retired couple experiences the highest degree of marital satisfaction since the first family stage, when they had no children.
  H. Despite high divorce rates, most marriages end with the death of a spouse.
    1. Widowhood is often associated with a significant decline in income, plunging the grieving spouse into financial crisis and hardship, particularly for poorer families.
    2. Recovering from the loss of a spouse is often difficult and prolonged.
    3. A large number of elderly men and women live together without remarrying.
    4. For many widows, widowhood lasts the rest of their lives.

VII. ENDURING MARRIAGES

  A. Long-term marriages may be divided into three categories: (1) couples who are happily in love, (2) unhappy couples who stay together out of habit or fear, and (3) couples who are neither happy nor unhappy.
    1. Researchers have found that approximately 20 percent of long term marriages were very happy, while 20 percent were very unhappy.
    2. Little correlation exists between happy marriages and stabile marriages.

VIII. FEATURES AND READINGS

  A. *Other Places... Other Times* describes the age set system of the Masai peoples of Kenya and Tanzania.
    1. Anthropologists and historians have found that persons in other cultures are not seen as "individuals" as we in the U.S.A. define the term.
    2. In most traditional societies, a person's identity is a clan identity.
    3. The age set system created fixed group identities and life stages, through which men of the age set moved together.
  B. *Understanding Yourself: Marital Satisfaction* reprints one widely used measure of marital satisfaction: Spanier's Dyadic Adjustment Scale.
  C. In *Perspective: Examining Marital Satisfaction* gives two main explanations for the decline in marital satisfaction which typically occurs soon after marriage.
    1. Marriages seem to give the most satisfaction when there are no children present.
    2. A second explanation for the decline in marital satisfaction is the duration-of-marriage effect: The accumulation over time of various factors such as unresolved conflicts, poor communication, grievances, role overload, heavy work schedules, and child-rearing responsibilities that cause marital disenchantment.
    3. Other important ingredients in marital satisfaction are social factors, psycological factors, attitudes toward gender and marital roles and expressiveness.
    4. Marital satisfaction is not static: It fluctuates over time.

Chapter 9

TEST YOUR COMPREHENSION

Below is a chart illustrating the eight stages of the family life cycle. (It should be noted that this model tends to assume children, which is certainly not always the case.) Complete the chart using readings, lecture and personal experience.

Chart 9

CHART OF THE FAMILY LIFE CYCLE	
Stage	**Description of Stage**
Beginning Families	
Childbearing Families	
Families with Preschool Children	
Families with School Children	
Families with Adolescents	
Families as Launching Centers	
Families in the Middle Years	
Aging Families	

Marriage as a Process: Family Life Cycle

SELF QUIZZES

How well do you know this material? Test your understanding of the reading assignments by answering the following sample questions.

PART I - Multiple Choice: Choose the most correct response.

_d_ 1. The developmental stage of the human life cycle which involves autonomy versus shame and doubt occurs during
   a. early childhood.
   b. adolescence.
   c. young adulthood.
   d. toddlerhood.
   e. infancy.

_c_ 2. The family life cycle
   a. involves six stages.
   b. encompasses single-parent and remarried families.
   c. uses a developmental framework to look at how families change over time.
   d. is sensitive to persons with disabilities.
   e. all of the above

_d_ 3. Factors which have been identified by researchers as associated with an engaged couple's eventual marital satisfaction include all but which one of the following?
   a. ability to resolve conflict constructively
   b. agreement on ethical issues
   c. communication patterns
   d. sexual pleasure
   e. balanced individual and couple leisure activities

_d_ 4. Which of the following is not a function of engagement?
   a. It prepares the couple to begin thinking about issues such as money, religion, in-laws.
   b. It helps define the goal of the couple's relationship.
   c. It increases involvement with future in-laws.
   d. It automatically means that a couple will become sexual.
   e. It provides the couple with a time period in which to strengthen themselves as a couple.

Chapter 9

___e___ 5. Identity bargaining involves
   a. each partner taking their idealized pictures of marriage and adjusting them to fit the reality of their own marriage.
   b. each partner having a clear identification with the role he or she is performing.
   c. each partner treating the other as if he or she fulfills the appropriate role.
   d. two people negotiating changes in each other's role.
   e. all of the above

___c___ 6. Which of the following is not a part of establishing boundaries with families of orientation?
   a. negotiating a different relationship with their families of orientation
   b. deciding how much influence their families of orientation may have
   c. maintaining primary loyalties with the family of orientation
   d. breaking habits of being subordinate
   e. forming a family that is interdependent rather than totally dependent or independent

___a___ 7. Which is not characteristic of the impact of children upon the marriage?
   a. Seventy-five percent of women quit working to attend to child rearing responsibilities.
   b. Working women must arrange child care and juggle employment responsibilities when children are sick.
   c. Non-employed mothers experience isolation and may face an identity crisis.
   d. The husband and wife tend to grow apart at this period.
   e. Marital satisfaction typically declines with the arrival of the first child.

___e___ 8. Families with adolescents
   a. may experience increased family conflict.
   b. must establish qualitatively different boundaries than families with younger children.
   c. ~~experience significant weakening of parental bonds.~~
   d. require considerable reorganization on the part of parents.
   e. all but c

___d___ 9. During later life,
   a. most persons between 65 and 74 years of age are widowed.
   b. the elderly are often neglected and isolated.
   c. couples experience a decline in marital satisfaction as they age.
   d. couples have their highest discretionary spending power.
   e. elderly men and women seldom live together without remarrying.

Marriage as a Process: Family Life Cycle

___e___ 10. Intermittent extended families
   a. are most likely to occur during middle-age marriages.
   b. are families that send money to other relatives in times of need.
   c. tend to remain expanded after the crisis is over.
   d. are not linked to any particular type of family background.
   e. "share and care" when younger or older relatives are in need or crisis.

___d___ 11. Grandparents responsible for parenting of their grandchildren are referred to as
   a. companionate grandparents.
   b. geographically remote grandparents.
   c. emotionally remote grandparents.
   d. involved grandparents.
   e. all of the above

___d___ 12. Contemporary grandparents
   a. fit the image of the lonely frail elderly person in a rocking chair.
   b. are less involved with grandchildren than previous generations.
   c. usually do not maintain important emotional roles in American families.
   d. frequently act as a stabilizing force for their children and grandchildren during divorce.
   e. are usually not involved in daily care of grandchildren.

___d___ 13. Which of the following is not characteristic of retirement?
   a. People highly identified with careers lose a major activity through which they define themselves.
   b. Retiring women retain important roles as wife, mother and homemaker, and still feel a sense of role fulfillment.
   c. If a woman has been a full-time homemaker, she probably doesn't retire from that role at all.
   d. The marriage is likely to suffer significant stress and unhappiness with the husband "underfoot."
   e. All of the above are characteristic of retirement.

___b___ 14. Which of the following is true regarding widowhood?
   a. Most marriages end in divorce rather than death.
   b. Recovering from the loss of a spouse is often difficult and prolonged.
   c. Widows who had bad marriages think of remarrying more often than those who had good marriages.
   d. Widows are usually financially secure.
   e. All of the above are false.

Chapter 9

___E___ 15. Which of the following is not true regarding enduring marriages?
   a. Approximately 20 percent of long-term marriages involve couples who are very happy.
   b. Approximately 20 percent of long-term marriages involve couples who are very unhappy.
   c. There is a strong correlation between marital happiness and stability.
   d. The quality of marital relationship appears to show continuity over the years.
   e. All of the above are true.

___E___ 16. Which of the following was not discussed in the text as being related to marital satisfaction.
   a. duration-of marriage effect
   b. expressiveness
   c. sexual satisfaction
   d. children
   e. All of the above were discussed.

PART II - True/False

___F___ 1. Marriage usually changes a person for the better.

___T___ 2. Age at marriage is a factor in the likelihood of divorce.

___T___ 3. Wedding traditions come from diverse sources such as ancient Egypt, Greece and Rome as well as the Middle Ages.

___T___ 4. Engagement is a time when people deal with anxiety and regret.

___F___ 5. Cohabitation rates in the United States have remained fairly constant, with little increase from previous years.

___F___ 6. The duration-of-marriage effect refers to the fact that stable marriages are usually happy marriages.

___F___ 7. Studies find that the happiest time in a marriage is after the birth of the first child.

___F___ 8. Research indicates that for most women the "empty nest" leads to depression and identity crisis.

___F___ 9. The number of adult children living at home is decreasing as more young people are seeking their own autonomous lifestyles.

Marriage as a Process: Family Life Cycle

F 10. Remote grandparents are usually people who are emotionally cold and reserved.

T 11. Despite the growing pains accompanying adolescence, parental bonds generally remain strong.

T 12. Intermittent extended families are most likely to occur in later-life families.

F 13. The most important factor in determining the amount of interaction between grandparents and grandchildren is health.

F 14. In general, couples who have experienced marital difficulty during their working years will experience improved marital relationships during retirement.

T 15. There is little correlation between marital happiness and marital stability.

F 16. The Dyadic Adjustment Scale measures marital stability.

PART II - Matching

Match the life stages, according to Erikson, with the developmental task which accompanies it.

d 1. Infancy
g 2. Toddler
h 3. Early Childhood
a 4. School Age
b 5. Adolescence
c 6. Young Adulthood
f 7. Adulthood
e 8. Maturity

a. industry vs. inferiority — School age
b. identity vs. role diffusion — Adolescence
c. intimacy vs. isolation — Young Adulthood
d. trust vs. mistrust — Infancy
e. integrity vs. despair — Maturity
f. generativity vs. self-absorption — Adulthood
g. autonomy vs. shame and doubt — Toddler
h. initiative vs. guilt — Early childhood

149

Chapter 9

DISCUSS BRIEFLY

1. How do traditional images of grandparenting compare with research findings about grandparenting?

2. What are some social changes which have impacted the traditional intact family life cycle? Why is recognition/acceptance of these changes important?

3. How does the failure to successfully complete the marital tasks of a newly married couple lead to the "duration of marriage effect?"

4. Because marriage and families are developmentally influenced, they are constantly changing. Discuss the importance of flexibility in both types of relationships. Cite examples of how rigidity in relationships may be harmful.

Marriage as a Process: Family Life Cycle

MINI-ASSIGNMENT I

If you were marrying (remarrying) now and decided to have a marriage contract, what specifics do you feel are important to include? (You might want to include areas such as name changes, children, sexual exclusiveness, division of chores, economic responsibility, etc. Also, see the next section for ideas.)

Do you think such a contract is a good idea? Why or why not?

FOR YOURSELF

The following is a list of questions people about to be married or establishing a long-term relationship may wish to consider and discuss with each other:

1. What kind of a commitment does a marriage entail? What happens when things don't seem to be working out (and this will happen!)? What are your thoughts on counseling, separation, divorce?

2. What are the rights of the man and woman involved? What role expectations do you have? What are the responsibilities of the husband? The wife? Who should be responsible for housework, cooking, finances, making decisions? Should someone have the right to make the final decisions alone?

Chapter 9

3. Does marriage involve sexual exclusiveness? How threatening would outside relationships, sexual and/or nonsexual, be for you and your partner?

4. Are children desired, and if so, how many? What is the role of children? What do they mean to you? Why should you have them? What kind of discipline do you prefer? Who is primarily responsible for discipline, day-to-day care, religious or moral training, sex education, and the instilling of values?

5. Who will be the "breadwinner", or will there be two? What is the view of the wife working? If the woman does work, is her job secondary to the man's? Is it a career or a job? Where does the money she earns working go? When/if children are born, should she work?

6. What is the role of religion in the marriage? How do you feel about your personal beliefs? What beliefs do each of you hold? Is religion necessary in your life? How do you feel about regular church attendance?

7. How do you feel about your future in-laws? How close are each of you to your own parents and siblings? What expectations of interaction do you have for the future with family members?

8. How is money to be spent? Are you spenders or savers, or both? Does one seem more important than the other? Do you have things you want to have "someday"? What things? Do you envision purchasing a home someday? What lifestyle do you predict for yourselves in ten years? Twenty years? When children are gone? Who manages money?

9. How do you envision your home life? Quiet and comfortable? Active and hectic? Is travel important? An active social life?

10. How do you see your home as a reflection of your individual personalities? Uncluttered and modern, or homey and traditional? A mixture? How important is this to each of you?

KEY TO SELF QUIZZES

Multiple Choice						True/False						Matching	
1.	d	7.	a	13.	d	1.	F	7.	F	13.	F	1.	d
2.	c	8.	e	14.	b	2.	T	8.	F	14.	F	2.	g
3.	d	9.	d	15.	c	3.	T	9.	F	15.	T	3.	h
4.	d	10.	e	16.	c	4.	T	10.	F	16.	F	4.	a
5.	e	11.	d			5.	F	11.	T			5.	b
6.	c	12.	d			6.	F	12.	T			6.	c
												7.	f
												8.	e

SUGGESTED READINGS

For related readings, see page 323 in the text.

CHAPTER 10

Parents and Children

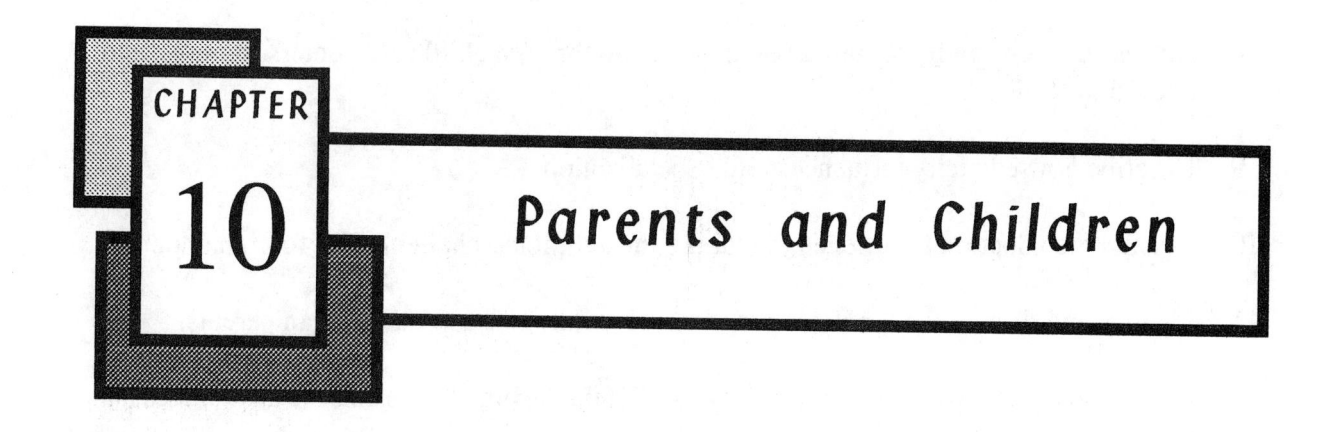

MAIN FOCUS

Chapter Ten explores issues related to the choice and timing of parenthood; theories of child socialization; children's developmental needs; parenting and parental needs; important child socializers; the issues of diverse families; and the styles and strategies of childrearing.

GOALS OF THIS CHAPTER

To demonstrate mastery of this chapter, you should be able to:

1. Formulate the issues to be considered regarding having a child and discuss the reason some people choose to have "childfree" marriages or defer parenthood.

2. Discuss the transition process which transpires when a child is first born, how the birth of a child affects the roles of both mother and father, and ways of dealing with the stress of transition to parenthood.

3. Understand and describe the basic theories of child socialization.

4. Explain children's developmental needs in terms of both nature and nurture.

5. Define high self-esteem and low self-esteem and discuss factors influencing self-esteem and ways to foster self-esteem.

6. Discuss the psychosexual development of children in the family context.

7. Understand the roles of motherhood and fatherhood as well as parental needs.

Chapter 10

8. Discuss the non-family socialization of the child through child care centers, schools, and television media.

9. Describe how ethnicity influences child socialization.

10. Discuss the adoption experience, including the additional challenges of foreign adoption.

11. Identify and discuss the family experiences of children of gay and lesbian parents.

12. Identify and describe the three basic styles of childrearing, discussing the advantages and disadvantages of each style.

13. Describe some of the childrearing strategies suggested by experts.

KEY TERMS AND IDEAS

The following terms, ideas, and concepts are listed in the order in which they appear in Chapter Ten. Be sure that you understand and can define each of the following:

child-free marriage	social learning theory
psychoanalytic theory	cognitive developmental theory
id	assimilation
superego	accommodation
ego	developmental systems approach
neurosis	attachment
psychosocial theory	authoritarian child rearing
behaviorism	permissive child rearing
reinforcement	authoritative child rearing
operant conditioning	

CHAPTER TEN OUTLINE

I. INTRODUCTION
   A. The family is the nursery of humanity: Traits that make us human are first developed within the family or its surrogate.
   B. Although today's families come in a variety of forms, they all seek to fulfill the needs of their members.

II. SHOULD WE OR SHOULDN'T WE? CHOOSING TO HAVE CHILDREN
   A. Parenthood can now be a matter of choice due to widespread use of birth control.
      1. The U.S. birthrate has fallen to an average of two children per marriage.
      2. Fertility rates vary considerable by race and ethnicity.

3. Cultural, social, and economic factors play a significant part in influencing the number of children a family has.
B. **Child-free marriages** are those in which the couples do not choose to have children; these couples are no longer seen as lacking something essential.
    1. Advantages for delaying parenthood include giving the parents a chance to complete their education, to build their careers, and to firmly establish their own relationship.
    2. Parents who have had a chance to establish themselves financially will be better able to bear the economic burdens of childrearing.
C. Factors contributing to deferred parenting include: more career and lifestyle options for single women; marriage and reproduction are no longer economic or social necessities; people may take longer to find the "right" mate or wait for the "right" time; and increasingly effective birth control.
    1. Deferred parenthood is in part due to later marriages, the fact that marriage and parenting are not economic necessities and the increased availability of birth control.
    2. It also allows parents time to become better financially established and emotionally secure.
D. The transition to parenthood signifies adulthood—the final, irreversible end of youthful roles.
    1. The abrupt transition to parenthood may create considerable stress; parents take on parental roles literally overnight, and the job goes on without relief around the clock.
    2. Overall, mothers seem to experience greater stress than fathers.
    3. Although a couple may have an egalitarian marriage before the birth of the first child, the marriage usually becomes more traditional once a child is born.
    4. Multiple role demands are the greatest source of stress for mothers.
    5. Couples experience less stress if they already have a strong relationship, use open communication, agree upon family planning, and originally had a strong desire for children.
    6. Ventura suggests that families can be assisted by improved healthcare and support in three areas: community support, coordination of care, and anticipatory care and problem solving.

III. THEORIES OF CHILD SOCIALIZATION
   A. Psychological theories focus on the role of the mind, particularly the subconscious mind.
       1. Sigmund Freud's **psychoanalytic theory** emphasizes unconscious mental processes in the development of personality.
           a. Freud's theory proposes that we are driven by instinct to seek pleasure, especially sexual pleasure.
           b. The **id** is the part of personality that seeks pleasure.
           c. The **superego** is the conscience

Chapter 10

      d.   The **ego** mediates between the id and the constraints of society.
      e.   According to Freud, too much restraint leads to repression and the development of **neuroses** (psychological disorders characterized by anxiety, phobia, etc.).
      f.   Freud viewed parents as mainly responsible for the child's psychological development.
   2.   In the **psychosocial theory** of development, Erik Erikson's life-cycle stages are each centered around a specific emotional concern based on individual biological influences and external sociocultural expectations and actions.

B. Learning theory emphasizes the aspects of behavior that are acquired rather than instinctual.
   1.   **Behaviorism**, developed by John B. Watson and B. F. Skinner, explains human behaviors on the basis of what can be observed.
      a.   Behaviorists reject the idea of hidden drives.
      b.   **Reinforcement** explains how behaviors may be increased or decreased.
      c.   **Operant conditioning** is the process of increasing the frequency of a behavior by reinforcement.
   2.   **Social learning theory** explains the role of thinking, or cognition, in learning.
      a.   Human nature is formed by the interactions of culture, society, and the family and the inner qualities of the individual.
      b.   Social learning theory draws from the ideas of behavioral psychology but adds to it the individual's innate ability to think and make choices to change his or her environment.

C. **Cognitive developmental theory**, developed by Jean Piaget, focuses on the discrete stages of development of the brain and nervous system which infants and children pass through at about the same time.
   1.   Children develop their cognitive abilities through interaction with the world and adaptation to their environment.
   2.   Children adapt by either **assimilation**, which involves adapting new information to fit their current understanding, or
   3.   **Accommodation**, which is adjusting their framework of understanding to fit in new experience.

D. The **developmental systems approach** views child growth and development as taking place within a complex and changing family system that both influences and is influenced by the child; the family system is nested within larger community and national systems.
   1.   Parents socialize children while children fulfill a socializing function for adults: Social and psychological development are lifelong processes.
   2.   Siblings influence one another in various ways making birth order and spacing important aspects of child socialization.

E. The sense of self or the "I" within each person is a crucial aspect of human personality.

Parents and Children

IV. CHILDREN'S DEVELOPMENTAL NEEDS
   A. A biological determinist believes that much of human behavior is guided by genetic makeup, physiological maturation, and neurological functioning.
   B. **Attachment** is the bonding between an infant and his or her primary caregiver(s).
      1. Babies signal their needs by gazing, crying, and smiling: They bond with people who are most responsive to them.
      2. Ainsworth discovered three patterns of infant attachment: secure, anxious/ambivalent, and anxious/avoidant.
   C. Individual temperament also influences a child's development: Parents who are sensitive to a child's unique temperament are better able to understand and influence the child.
   D. According to Konner, basic needs include:
      1. adequate prenatal nutrition and care,
      2. appropriate stimulation and care of newborn,
      3. the formation of at least one close attachment during the first five years,
      4. support for the family, including child care when a parent or parents must work,
      5. protection from illness,
      6. freedom from physical and sexual abuse,
      7. supportive friends, both adults and children,
      8. respect for the child's individuality and the presentation of appropriate challenges leading to competence,
      9. safe, nurturing, and challenging schooling,
      10. an adolescence "free of pressure to grow up too fast, yet respectful of natural biological transformations," and
      11. protection from premature parenthood.
   E. High self-esteem is essential for growth in relationships, creativity and productivity.
      1. Low self-esteem creates feelings of powerless ness, poor ability to cope, low tolerance for differences and difficulties, inability to accept responsibility, and impaired emotional responsiveness.
      2. Clemes and Bean describe four conditions necessary for developing and maintaining high self-esteem.
         a. a sense of connectedness
         b. a sense of uniqueness
         c. a sense of power
         d. a sense of models
      3. Timely, honest, specific feedback, rather than superficial praise, increases a child's self-esteem.
   F. Psychosexual development in the family context includes attitudes, behaviors, and values with respect to nudity, physical contacts, and physical affection.
      1. All children and adults need a certain amount of freely given physical affection from those they love.
      2. Children should be told, in a nonthreatening way, what kind of touching by adults is "good" and what kind is "bad."
      3. Children need to know that if they are sexually abused, it is not their fault: They need to know that they can tell about it and still be worthy of love.

Chapter 10

V. PARENTHOOD
   A. Many women see their destiny as motherhood; researchers have not found any instinctual motivation for having children among humans, but there are many social motives impelling women to become mothers.
      1. While most women choose motherhood, many feel ambivalent about motherhood.
      2. The expectation that motherhood comes naturally can be frustrating and guilt producing: New mothers face tremendous pressures.
      3. Patience with oneself, support from family and friends, and a partner who takes equal responsibility can help alleviate stress and bring greater joy to motherhood.
   B. The father's traditional roles of provider and protector are instrumental; they satisfy the family's economic and physical needs.
      1. The mother's traditional roles are expressive; she gives emotional support for her family.
      2. At present, the lines between these roles are becoming increasingly blurred because of economic pressures and new social expectations.
      3. Many fathers trend to be differently involved with their male and female children.
      4. Most men today compare themselves favorably with their own fathers in both quality and quantity on involvement with their children.
   C. Important needs of parents during the childrearing years include personal developmental needs and the need to maintain marital satisfaction.

VI. OTHER IMPORTANT CHILD SOCIALIZERS
   A. With the rise of single parent families and maternal employment, children are increasingly socialized by influences outside the immediate family.
   B Supplementary child care is a crucial issue for today's parents of young children.
      1. Most experts agree that the ideal environment for raising a child is in the home with the parents and family: Since this ideal is not often possible, the role of day care needs to be considered.
      2. The results of research on daycare are mixed: High-quality care, given by sensitive, responsive, and stimulating caregivers in a safe and low teacher-to-student ratio, can facilitate the development of positive social qualities, consideration, and independence.
      3. A national study found that children have a far greater likelihood of being sexually abused by a father, step-father, or other relative than by a daycare worker.
      4. In addition to cleanliness, comfort, good food, and a safe environment, parents should be familiar with state licensure requirements, check references, and observe caregiver with their children.
      5. The American Academy of Child and Adolescent Psychiatry offer nine suggestions for parents seeking day-care services.
      6. The United States is one of the few industrialized nations that does not have a comprehensive national daycare policy.

C. Schools play an important role in children's socialization, yet in most areas there is less and less money allocated for education or the maintenance of schools.
   1. Success in schools is very highly correlated with socioeconomic status.
   2. Entering kindergarten marks the beginning of profound changes in a child's life.
   3. Interactions with teachers and performance of the tasks of learning will become the basis for the child's evaluation of his or her self-worth.
D. Starting around the age of four, most children watch an average of two to three hours of television each day: No other activity, except sleep, occupies more time for children and adolescents.
   1. Time spent watching television limits children's time for other pursuits.
   2. Television has been implicated in a number of individual, familial, and societal disorders.
   3. The use of television as a baby-sitter underscores a dilemma for many parents: They disapprove of what their children see, but they want to have time to themselves.
   4. The controversial age-based programming rating divides programs into six categories.

VII. Issues of Diverse Families
  A. Our families are the key to the transmission of ethnic identification.
   1. A child's ethnic background effects how he or she is socialized.
   2. Groups with minority status in the U.S. usually emphasize education as the means for their children to achieve success.
   3. Discrimination and prejudice shape the lives of many American children; their parents may try to prepare their children for the harsh realities of life beyond the family and immediate community.
  B. Slightly more than two percent of the American population is adopted.
   1. Although tens of thousands of parents and potential parents are currently waiting to adopt, there is a shortage of available healthy babies.
      a. The costs for open adoptions tend to run between $6,000 and more than $20,000, with some paying up to $100,000.
      b. Each state has its own adoption laws which can either constrain or support open adoptions.
      c. The trend is toward open adoption in which there is contact between the adoptive family and birth parents.
   2. Foreign adoptions are also increasingly favored: Approximately 15 percent of U.S. adoptions are of children born outside the U.S.
      a. Families with children from other cultures face unique challenges in addition to those faced by other adoptive families.
      b. Adoptive parents usually receive little information about the foreign birth parents and have no opportunities for continued contact.
      c. Older children often feel rootlessness and loss.

3. Adoptive families face unique problems: the stresses of infertility; disappointments and frustration during the wait; the fact that they may have spent all their savings; and insensitivity and prejudice.
C. Researchers believe that the number families with at least one gay parent ranges from 6 to 14 million: Most of these parents are, or have been, married.
1. Lesbian mothers share many similarities with heterosexual single mothers; their life-styles, childrearing practices, and general demographic data are strikingly similar.
   a. A summary of studies on gay parenting concluded that the children are just as well-adjusted and no more likely to be gay themselves.
   b. Many lesbians fear losing their children in custody battles: The courts have increasingly taken the position that the sexual orientation of mothers is not an issue if the children are well-cared-for and well-adjusted.
   c. The children of lesbians tend to have difficulty in accepting their mother's identity because there is no community of support.
   d. Some lesbians, especially those in committed relationships, are choosing to create families through artificial insemination.
2. Studies of gay fathers indicate that "being gay is compatible with effective parenting."
   a. One study of gay fathers concluded that: most gay fathers have positive relationships with their children; the father's sexual orientation is relatively unimportant to the relationship with their children; and gay fathers endeavor to create stable home environments.
   b. Children of gay fathers often worry about what others may think of their fathers.
3. Heterosexual fears about the parenting abilities of lesbians and gays are unwarranted.
   a. Fears of the sexual abuse of children by gay parents or their partners are completely unsubstantiated.
   b. Fears about gay parents rejecting children of the other sex also seem to be unfounded.

VIII. STYLES AND STRATEGIES OF CHILDREARING
A. A parent's approach to training, teaching, nurturing, and helping a child varies according to cultural influences, the parent's personality and basic attitudes toward children and childrearing, and the role model that the parent presents to the child.
B. The three basic styles of childrearing are authoritarian, permissive, and authoritative.
1. **Authoritarian parents** typically require absolute obedience.
   a. The parents' maintaining control is of first importance.
   b. Working-class families tend to be more authoritarian than middle-class families.
2. **Permissive parents** value the child's freedom of expression and autonomy; they rely on reasoning and explanations: Permissive attitudes are more popular in middle-class families.

3. **Authoritative parents** rely on positive reinforcements and infrequent use of punishment: They direct the child in a manner that shows awareness of his or her feelings and capabilities.
C. Many contemporary parents rely on the advice of experts even when it conflicts with their own experience.
D. Childrearing strategies currently taught or endorsed by experts include:
1. mutual respect between children and parents,
2. consistency and clarity,
3. logical consequences,
4. open communication including active listening and the use of I-messages,
5. no physical punishment; and
6. behavior modification

IX. FEATURES AND READINGS
A. The reading *Am I Parent Material?* is designed to stimulate ideas about the decision to have and raise children.
B. *Other Places ... Other Times, Families of Mexican Origin* discusses the socialization of children in Mexican Families.
1. Boys and girls are socialized differently.
2. Assimilation and acculturation have impacted socialization.
3. The extended family is the mainstay: Family is central and is more important than individualism
C. *Other Places ... Other Times, The Rights of Biological Parents Versus Social Parents in Cross-Cultural Perspective: The Nuer* explores how the Nuer of the Sudan in East Africa transfer cattle related to child custody.
D. In *You and Your Well-Being, Dealing With Parent Burnout* eight suggestions are given for dealing with the problems of parent burnout.
E. In *You and Your Well-Being, Children and Television: What's the Verdict?* the authors suggest that television is not inherently harmful if it is used wisely: Measures that parents can take to minimize the negative effects of television are offered.
F. The *Perspective: Raising Children to Be Prejudice Free* offers several guidelines to help parents foster an appreciation of cultural differences in their children.

Chapter 10

TEST YOUR COMPREHENSION

The following chart lists the leading theories of child socialization. Give the name of the person or persons responsible for each theory, and then describe in detail the important concepts for each theory.

Chart 10

THEORIES OF CHILD SOCIALIZATION		
Theory	**Person**	**Concepts**
Psychoanalytic Theory		
Psychosocial Theory		
Behaviorism		
Social Learning Theory		
Cognitive Developmental Theory		
Developmental Systems Approach		

SELF QUIZZES

How well do you know this material? Test your understanding of the reading assignments by answering the following sample questions.

PART I - Multiple Choice: Choose the most correct response.

b 1. Which of the following is not generally true of couples who choose to be child-free?
   a. Approximately 9.3 percent of American women between the ages of 18 and 34 do not plan to have children.
   b. Couples who choose to be child-free should best be viewed as lacking something essential for personal fulfillment.
   c. Women who choose to be child-free are generally well-educated and career-oriented.
   d. The decision is generally realized gradually.
   e. These families have a higher degree of marital adjustment and satisfaction than couples with children.

a 2. Childbirth is frequently being deferred because
   a. the average age of marriage has increased.
   b. there are more options available to single women than in the past.
   c. of increasingly effective birth control options.
   d. older parents may be more financially and emotionally prepared for parenting.
   e. all of the above

e 3. Which of the following should one consider when deciding to have a child or another child?
   a. What are the real motives for having a child?
   b. Am I under social or family pressure to have a child?
   c. What would an additional family member do to the present relationship?
   d. What are the financial responsibilities a child would add?
   e. all of the above

b 4. All but which one of the following is true regarding the transition to parenthood?
   a. Many parents express concern about their ability to meet all the responsibilities of child rearing.
   b. Couples with egalitarian marriages are likely to continue egalitarian roles in parenting.
   c. Parenthood signifies adulthood--the final, irreversible end of youthful roles.
   d. Overall, mothers experience greater stress than fathers.
   e. All of the above are true.

Chapter 10

___c___ 5. _____ stresses the importance of unconscious mental processes and the stages of psychosexual development.
   a. Psychosocial theory
   b. Behaviorism
   c. Psychoanalytic theory
   d. Social learning theory
   e. Cognitive developmental theory

___c___ 6. Piaget's cognitive developmental theory proposes that children adjust to their surroundings through _____ and _____.
   a. operant conditioning and reinforcement
   b. individual biological influences and external sociocultural expectations
   c. assimilation and adaptation
   d. cognition and socialization
   e. unconscious mental processes and instinct

___e___ 7. According to Clemes & Bean, developing and maintaining high self-esteem requires a sense of _____.
   a. connectedness
   b. uniqueness
   c. power
   d. models
   e. all of the above

___d___ 8. Basic needs of children in terms of psychosexual development include each of the following except
   a. acceptance and respect for the child's body and nudity.
   b. a feeling that they are in charge of their own bodies.
   c. a certain amount of freely given physical affection.
   d. mild punishment when interrupting parent's love making.
   e. All of the above are psychosexual developmental needs.

___a___ 9. Many women become mothers through
   a. no conscious choice.
   b. instinctual motives.
   c. social motives.
   d. a sense of destiny.
   e. all but b

Parents and Children

___a___ 10. Which of the following is not true of fathering?
   a. It tends to mean the same as mothering.
   b. It can refer to as little as the male participation in the fertilization process.
   c. Parenting has been a more peripheral role for men, while breadwinning has been the principal role.
   d. The fathering role has not traditionally involved nurturing.
   e. Fathers tend to be differently involved with their male and female children.

___c___ 11. All but which one of the following statements is true regarding child care?
   a. Children have a greater likelihood of being sexually abused by a relative than a day care worker.
   b. Those who need child care the most can least afford it.
   c. Overall, results of research on day care strongly indicate that child care has a positive effect on children.
   d. High-quality child care can facilitate the development of positive social qualities.
   e. All of the above are true.

___a___ 12. Ethnicity
   a. affects child socialization.
   b. has only subjective components.
   c. is fixed and unchanging.
   d. is of little concern to parents of minority children.
   e. both a and c

___e___ 13. Families headed by lesbian and gay men
   a. generally experience the same joys and pains as those headed by heterosexual women and men.
   b. are likely to face insensitivity from society.
   c. are likely to face discrimination from society.
   d. are just as well adjusted as families headed by heterosexual parents.
   e. all of the above

___a___ 14. An attitude towards childrearing which typically requires absolute obedience is likely to be held by
   a. authoritarian parents.
   b. middle-class parents.
   c. permissive parents.
   d. authoritative parents.
   e. all of the above

Chapter 10

_a_ 15. Techniques of child-rearing currently endorsed by experts include each of the following except
   a. mild physical punishment.
   b. behavior modification.
   c. logical consequences.
   d. consistency and clarity.
   e. All of the above are endorsed.

_d_ 16. Families of Mexican origin
   a. value individualism.
   b. pay little attention to gender when raising children.
   c. tend to be authoritarian.
   d. have an extended family structure that serves as a vital link between family and community.
   e. both a and d

_e_ 17. It is important for parents to
   a. learn how to develop self-esteem.
   b. reward themselves for doing a good job.
   c. define personal goals.
   d. forgive themselves when they are less than perfect.
   e. all of the above

_d_ 18. Each of the following is true regarding children and television except
   a. the more television children watch, the more they subscribe to male-female stereotypes.
   b. music television videos are probably the most guilty of presenting negative attitudes toward women.
   c. children's viewing of violent programs has been shown to cause difficulties with the ability to stay focused on a task.
   d. television is inherently harmful.
   e. neither music nor television is a danger for a child whose life is happy and healthy.

PART II - True/False

_F_ 1. Fertility rates vary little by ethnicity.

_T_ 2. Many studies of child-free marriages indicate a higher degree of marital adjustment than is found among couples with children.

Parents and Children

__F__ 3. Marriages which were egalitarian before child-birth will tend to continue to be egalitarian after the baby is born.

__T__ 4. The father's emotional and physical support are important factors in maintaining marital quality during the transition to parenthood.

__F__ 5. Freud emphasized the effects of society on the developing ego.

__F__ 6. According to Piaget, children vary considerably in terms of the ages at which they pass through the stages of cognitive development.

__T__ 7. The developmental systems approach proposes that the socialization process is a two-way process.

__F__ 8. A child's unique temperament is biologically based and not subject to external influences.

__F__ 9. A child's sense of power has little impact on his or her self-esteem level.

__T__ 10. Many women who are mothers made no conscious choice to have a baby, possibly because they didn't know that a choice was possible.

__T__ 11. Motherhood is instinctive.

__F__ 12. The traditional roles of the father are instrumental, while those of the mother are expressive, but the lines between these roles are becoming increasingly blurred.

__T__ 13. Research indicates that much more than mothers, fathers tend to be involved differently with their male and female children.

__T__ 14. In emphasizing children's needs, parents often fail to recognize the importance of their own needs as well.

__F__ 15. The emotional security and patterns for attachment of young children to parents seems to be significantly altered by group daycare experiences.

__F__ 16. Statistics show that sexual abuse by a babysitter or daycare worker is a major part of reported cases of child abuse.

__F__ 17. Children raised by lesbian or gay parents are much more likely to have a homosexual preference when they grow up.

__T__ 18. Involved, devoted and enthusiastic parents are those most likely to burnout.

Chapter 10

DISCUSS BRIEFLY

1. Discuss the evidence of a maternal instinct vs. socialization of females to become mothers.

2. What are some of the reasons and myths our culture gives us for having a baby?

3. What are some reasons we might have for choosing not to have a baby now (or maybe ever)?

Parents and Children

4. Discuss how parental upbringing and attitudes can influence a child's self-esteem level.

5. Discuss how a parent's own sexuality influences the psychosexual development of their child(ren).

SELF-DISCOVERY

What do you think are some valid reasons for deciding to have a baby? Do you (did you) want children?

Chapter 10

In your family of origin, what childrearing attitudes (authoritarian, permissive, or authoritative) predominated? How do you think these attitudes influenced your own development? What childrearing strategies do you lean toward the most?

KEY TO SELF QUIZZES

Multiple Choice

1. b   10. a
2. e   11. c
3. e   12. a
4. b   13. e
5. c   14. a
6. c   15. a
7. e   16. d
8. d   17. e
9. a   18. d

True/False

1. F   10. T
2. T   11. T
3. F   12. F
4. T   13. T
5. F   14. T
6. F   15. F
7. T   16. F
8. F   17. F
9. F   18. T

SUGGESTED READINGS

For related readings, see 363 in the text.

CHAPTER 11

Marriage, Work, and Economics

MAIN FOCUS

Chapter Eleven examines workplace-family linkages; employment and the family life-cycle; the families division of labor; women in the labor force; and dual earner marriages. It also looks at family issues in the workplace; unemployment; poverty; and family policy.

GOALS OF THIS CHAPTER

To demonstrate mastery of this chapter, you should be able to:

1. Explain the ramifications of employment taking precedence over family in the United States, particularly in terms of family work and paid work.

2. Understand the relationship of economics to the family cycle and the quality of family life.

3. Describe and give examples of work spillover, role overload and interrole conflict.

4. Understand and describe the stages of the traditional work/family model and the alternative work/family life cycle approaches.

5. Understand and explain the traditional division of labor and how it is currently being challenged by changing roles of men and women.

6. Describe and evaluate the homemaker role and discuss why it is frequently devalued by our culture.

7. Discuss the role of women in the labor force and how this effects the family.

Chapter 11

8. Evaluate the dynamics of the dual-career marriages and describe the advantages and disadvantages of such arrangements to the modern family.

9. Describe the economic forces which discriminate against women and the social changes being suggested which might improve this situation, including comparable worth.

10. Discuss how sexual harassment can be used at different levels to discriminate against individuals.

11. Discuss the lack of adequate child care, how this affects families, and ways which society could ease this burden.

12. Understand and describe aspects of economic distress on the family, including unemployment, underemployment, poverty, homelessness, and the role of public assistance.

13. Describe the myths surrounding welfare and the ways in which they are inaccurate.

14. Define family policy and discuss the importance of families in terms of United States governmental policies.

KEY TERMS AND IDEAS

The following columns of terms, ideas, and concepts are listed in the order that they appear in Chapter Eleven. Be sure that you understand and can define each of the following:

family work	occupational stratification
work spillover	self-care
role strain	economic distress
role overload	Aid to Families with Dependent Children (AFDC)
interrole conflict	wealth
homemaker role	feminization of poverty
sexual harassment	ghetto poor
hostile environment	family policy

CHAPTER ELEVEN OUTLINE

I. INTRODUCTION TO CHAPTER ELEVEN
   A. In the United States, employment takes precedence over family.
   B. Families are economic units bound together by emotional ties.
   C. **Family work** is work performed in the home without pay: Women perform the overwhelming majority of family work.
   D. Paid work helps shape the quality of family life: It affects time, roles, income, spending, leisure, even individual identities.
   E. Paid work is inflexible and regulates the family.

II. WORKPLACE/FAMILY LINKAGES
   A. Work-family conflict is often a painful source of stress.
   B. **Work spillover** — the effect of work on other aspects of our lives — is important in family life.
      1. Work is as much a part of marriage as love is.
      2. Work spillover particularly affects women: An employed women's leisure is rarely spontaneous.
   C. Work-related problems lead to **role strains**: Role strain is the most pressing problem for working families.
      1. **Role overload** occurs when the total prescribed activities are more than one individual can comfortably or adequately handle.
      2. **Interrole conflict** occurs when the expectations of the parent/spouse/worker roles are contradictory or require doing two things at once.

III. EMPLOYMENT AND THE FAMILY LIFE CYCLE
   A. Research indicates that women spend considerably more total time in work and family roles than do men.
   B. Much marital conflict in families today grows out of inequities between male and female work and family role expectations and experiences.
   C. The traditional-simultaneous work/family life cycle model characterizes families with the husband as the major economic provider and the wife as caregiver and responsible for household tasks.
      1. The establishment/novitiate stage occurs when men and women are beginning to establish their families and begin paid work.
      2. The new parents/early career stage involves role overload as an acute problem.
      3. The school-age/middle career stage has high interrole conflicts: Parental roles, more than spousal roles, contribute to role overload and interrole conflict.
      4. The postparental family/late career stage involves more free time, an empty nest, a renewing of spousal ties, and sometimes the caring for aged parents.

Chapter 11

D. The burden on women of combining work and family roles has resulted in alternative work/family cycle approaches.
  1. The sequential work/family role staging pattern involves the woman leaving work with the birth of the first child.
     a. In the conventional pattern, the woman quits her job and does not return.
     b. In the early interrupted pattern, the woman stops working early in her career to have children and resumes working later.
     c. In the later interrupted pattern, the woman first establishes her career, quits to have children, then returns.
     d. In the unstable pattern, the woman goes back and forth between full-time paid employment and homemaking according to economic need.
  2. The symmetrical work/family role allocation involves both the husband and the wife being employed, and both sharing family work.

IV. THE FAMILY'S DIVISION OF LABOR
  A. How families divide labor has a tremendous impact on how a family functions. In the traditional division of labor (the complementary model), the husband is expected to work outside the home for wages and the wife is expected to remain at home caring for children and maintaining the household.
    1. The traditional role of provider remains primary for men: men's family work involves household maintenance and repair, light construction, mowing the yard and "helping" their partner in household tasks.
    2. Because the **homemaker role** is performed by women and is unpaid, it has been denigrated.
    3. Women's family work is considerably more diverse than men's and permeates every aspect of family life.
    4. No matter what kind of work a woman does outside the home, there is seldom equality when it comes to housework.
    5. Ann Oakley described four characteristics of the homemaker role:
       a. its exclusive allocation to women,
       b. its association with economic dependence,
       c. its status as non-work, and
       d. its priority over other roles for women.
  B. Housework has the following characteristics:
    1. It tends to isolate a woman alone at home.
    2. It is unstructured, monotonous, and repetitive.
    3. The full-time homemaker role is restricted.
    4. Housework is autonomous.
    5. Homemakers work long days and nights.
    6. Homemaking can involve child rearing.
    7. It often involves role strain.
    8. Housework is unpaid.

III. WOMEN IN THE LABOR FORCE

A. Research indicates that women's employment tends to have positive rather than negative effects on marriage.

B. Between 1960 and 1995, the percentage of married women in the labor force almost doubled.

C. Factor's which influence a woman's decision to enter the labor force include financial considerations, social norms, self-fulfillment, and attitudes about employment and family.

D. Employment tends to improve women's emotional and physical well being as it decreases economic hardship and increased domestic support from her partner.

E. Women's employment patterns reflect their family and child-care responsibilities.
   1. Women tend to interrupt their job and career lives considerably more than men.
   2. A woman's decision to remain in the work force or withdraw from it during her childbearing and early child-rearing years is critical for her later work force activities.

VI. DUAL-EARNER MARRIAGES

A. Economic changes related to real wages, loss of manufacturing jobs, and the rise of a low-paying service economy have lead to a significant increase in dual-earner marriages.

B. Even though two-thirds of all married women held jobs in 1995, the great majority of them were employed in low-paying, low-status jobs.

C. In dual-career families, both the husband and wife have high achievement orientations, greater emphasis on gender equality, and a stronger desire to exercise their capabilities.

D. Marital satisfaction is tied to fair division of household labor: Dividing household work fairly may be a key to marital success.

E. Women's employment appears to have little impact on division of household labor: Employed women do twice as much housework as men.
   1. Working wives continue to have primary responsibility for running the family; they do, however, spend fewer hours on housework.
   2. The husband's contribution to household work does not significantly increase if the wife is employed.
   3. Men's participation in housework increases as they have less demanding jobs, as their wives' income rises, and as they are more expressive.
   4. Equitable division is not the same as equal division of housework: What is fair is determined differently by different couples.
   5. Men increasingly believe they should be more involved as fathers, but the shift in attitude toward child-rearing responsibilities is not reflected in their behavior.

F. Marital power of the wife increases with her occupational status and her income.

Chapter 11

    G. Role strain is a constant factor for employed women: Women generally make greater adjustments than men in dual-earner marriages
        1. Studies suggest a relationship between women's employment and the likelihood of divorce, except for African-American women.
        2. The effect of a wife's full time employment on a couples marital satisfaction is influenced by social class, the presence of children, and the husband's and wife's attitudes and commitments to her working.
    H. The three greatest social needs in a dual-career marriage are, (1) redefining gender roles to eliminate role overload for women, (2) providing adequate child-care facilities for working parents, and (3) restructuring the workplace to recognize the special needs of parents and families.
        1. Coping strategies include reorganizing the family system and reevaluating house hold expectations.
        2. Coping techniques may include hiring outside help, reducing employment hours, or working different shifts to facilitate child care.

VII. FAMILY ISSUES IN THE WORKPLACE

    A. Many workplace issues directly impact families: A women's earnings have a significant impact on family well-being.
        1. There is a significant earnings gap, with women making 75 cents for every dollar that men earn, resulting in many women being condemned to poverty and forced to accept welfare.
        2. Wage differentials are especially important to single women.
        3. Much of the earnings gap is the result of occupational differences, gender segregation, and women's tendency to interrupt their employment for family reasons.
    B. **Sexual harassment** is a mixture of sex and power of two types: (1) the abuse of power for sexual ends; and (2) the creation of a **hostile environment** in which someone acts in sexual ways so as to interfere with a person's performance by creating a hostile or offensive learning or work environment.
        1. Sexual harassment is illegal.
        2. Gender differences may contribute to sexual harassment:
            a. Men are less likely than women to view sexual activities as harassment.
            b. Men misinterpret women's friendliness as sexual interest.
            c. Men are more likely to perceive male-female relationships as adversarial.
        3. Power differences affect perception.
    C. **Occupational stratification** is the hierarchical ranking of jobs by income and status and is often related to gender and ethnicity.
        1. Almost one-third of African-Americans work in middle-class or professional occupations; the remainder are typically employed in declining industries or the lowest paying jobs.

2. Technological changes and foreign competition have crushed American manufacturing: These are industries where African Americans were once employed in well-paying jobs.
   3. Latinos (with the exception of Cuban Americans) tend to be concentrated in blue-collar or service work.
   4. Ethnic women tend to be employed in the lowest-paying of traditionally low-paying female jobs.
D. For women, inadequate child care is one of the major barriers to equal employment and educational opportunity.
   1. Child care is difficult to obtain and many parents are forced to constantly switch to find or maintain day care arrangements.
   2. The high cost of day care is harmful to poorer families and is a major force keeping mothers on welfare from working.
   3. Lack of child care prevents women from taking paid jobs, keeps them in part-time jobs, prevents them from seeking promotions, interferes with their work and may restrict further educational opportunity.
E. Due to lack of after-school programs or prohibiting costs, **self-care** is now a major form of child care with varying degrees of success and problems.
   1. Self-care exists in families of all economic classes.
   2. Research on self-care is sketchy and contradictory.
F. Although women make up a significant part of the work force, businesses have maintained an inflexible work environment.
   1. Most businesses are run as if every worker were male with a full-time wife at home; women make up a significant part of the work force and they do not have wives.
   2. Employees who feel supported by their employers in terms of family responsibilities are less likely to experience work-family role strain.
   3. Family-oriented employment policies could include flexible work schedules, job sharing, extended maternity/paternity leave and benefits, more personal leave days, corporate child-care, part-time benefits and flexible benefit programs.

VIII. Unemployment is a major source of stress for individuals, with its consequences spilling over into their families.
  A. **Economic distress** includes those aspects of economic life, such as unemployment, poverty and economic strain.
  B. The emotional and financial cost of unemployment to workers and their families is high.
  C. Resources for coping with unemployment include an individual's psychological disposition, a strong sense of self-esteem, and a feeling of mastery.
  D. Important coping behaviors include defining the meaning of the problem, problem solving, and managing emotions.

Chapter 11

IX. POVERTY
   A. The family and economy are intimately connected to each other, and economic inequality directly affects the well-being of America's disadvantaged families.
      1. Poverty is associated with marital and family stress, increased divorce rates, homelessness, low birth weight and infant deaths, poor health, depression, lowered life-expectancy and feelings of hopelessness and despair.
      2. Poverty has been increasing since 1981, primarily due to sharp swings in employment, an increase in single-parent families, and government cutbacks in assistance to low-income families.
   B. Spells of poverty tend to be temporary rather than permanent.
      1. These spells are frequently caused by changes in circumstances due to divorce, unemployment, illness, disability or death.
      2. Many who accept government assistance return to self-sufficiency within a year or two.
      3. Major factors related to the beginning and ending of spells of poverty are changes in income and changes in family composition.
      4. Welfare or government assistance is frequently in the form of **Aid to Families with Dependent Children (AFDC)**.
      5. Poverty spells are shorter if they begin with a decline in income than if they begin with a transition to single parenthood.
   C. There are vast disparities in income and **wealth** between White, African-American, Latino and other ethnic groups: Disparity in wealth exceeds disparity in income.
   D. Since 1979, the largest increase in the numbers of poor has been among the working poor because of low wages, occupational segregation and the rise in single-parent families; they cannot earn enough to raise themselves from poverty.
   E. Divorce and the increasing numbers of unmarried women with children have contributed to the **feminization of poverty**.
   F. The **ghetto poor**, inner city residents, primarily African-Americans and Latinos, along with the homeless have become deeply disturbing features of American life.
   G. The war on poverty of the 1960s has become the war on welfare of the 1980s and 1990s.
      1. Instead of viewing poverty as a structural feature of our society, we increasingly blame the poor for their poverty.
      2. Welfare has become a central issue in contemporary politics.

X. **Family policy** is a set of objectives concerning family well-being and the specific government measures designed to achieve those objectives.
   A. In examining America's priorities, it is clear that families are not a national priority; neither are women and children.
      1. None of the family support programs that require substantial monies, such as AFDC and WIC, are adequately funded.
      2. There is increasing interest in a systematic approach to family policy.

B. Elements which would support the family would affect health care, social welfare, education and the workplace
   1. Positive health care would include: guaranteed adequate health care for every citizen, prenatal and infant care for all mothers; adequate nutrition, immunizations and clinics for baby health; care for the aged, disabled and assistance for caregivers; comprehensive and realistic sex education; education about STDs; and drug and alcohol rehabilitation available to all who need them.
   2. Social welfare programs could include tax credits or income maintenance program for families; a basic standard of living for all children; child care for working or disabled parents; advocacy for those who are without power; attention to the problems of the homeless; and regulation of children's television to promote literacy, humane values and critical thinking.
   3. Education programs needed include: pre-school education; an emphasis on teaching basic skills and reaching all students; job skills, bilingual and multicultural programs; special education for the developmentally disabled; and adult education.
   4. Workplace reforms might include: paid parental leave for pregnancy and sick children; flexible work schedules for parents; increased minimum wage, fair employment for all; pay equity between men and women; affirmative action policies; corporate child-care programs; individual and family counseling services; and flexible benefits.

XI. READINGS AND FEATURES
   A. *Other Places ... Other Times: Industrialization 'Creates' the Traditional Family* traces the historical changes brought about in the family by the transition from self-sufficient farm family to wage-earning urban family.
   B. In *Understanding Yourself: The Division of Labor: A Marriage Contract* questions concerning the importance of marriage, job, division of housework, and child care are discussed.
   C. *The Perspective: Hungry and Homeless in America* discusses the ravages of poverty upon individuals and, increasingly, families with women and children.
      1. Poverty increased dramatically in the 1980s and increased substantially again in 1991 with a new recession.
      2. Housing prices skyrocketed while low-income rentals were decreasing, resulting in more than twice as many poor households as there is affordable housing.
      3. The traditional welfare system underwent relentless attack, reducing single mothers to homelessness.
      4. A commitment to ending homelessness would include (1) the development of low cost housing; (2) the establishment of child-care programs that allow homeless single mothers to work; (3) the provision of mental health support; and (4) the initiation of job training and education programs.

Chapter 11

TEST YOUR COMPREHENSION

On the next two pages is a chart illustrating the effects upon the family of the wife working outside the home. Your text has discussed many aspects of this. Complete the chart using readings, lecture and personal research.

Chart 10

ADVANTAGES AND DISADVANTAGES OF SINGLE - CAREER FAMILY
The psychological impact on the woman as homemaker: Positive   vs.   Negative
The psychological impact on the man as breadwinner: Positive   vs.   Negative
The economic effect of one-earner family in terms of: Financial benefits   vs.   Financial costs
The effects upon the children: Positive   vs.   Negative
The effects upon housekeeping chores: Positive   vs.   Negative

SINGLE - CAREER   VS.   DUAL - CAREER   MARRIAGES
DUAL - CAREER   FAMILY
The psychological impact on the woman who works: Positive   vs.   Negative
The psychological impact on the man whose wife works: Positive   vs.   Negative
The economic effect of dual-career family in terms of: Financial benefits   vs.   Financial costs
The effects upon the children: Positive   vs.   Negative
The effects upon housekeeping chores: Positive   vs.   Negative

Chapter 11

SELF QUIZZES

How well do you know this material? Test your understanding of the reading assignment by answering the following sample questions.

PART I - Multiple Choice: Choose the most correct response.

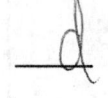

__d__ 1. Which of the following is not true regarding work and family life?
   a. When most people think of work, they tend to recognize only paid work.
   b. Work is as much a part of marriage as love.
   c. As a result of work spillover, an employed woman's leisure is rarely spontaneous.
   d. Most dual-earner couples divide housework nearly fifty-fifty.
   e. All of the above are true.

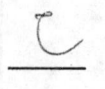

__c__ 2. Which of the following does not characterize role overload?
   a. Feeling overwhelmed by multiple responsibilities of parent/spouse/worker.
   b. The total prescribed activities are greater than an individual can handle.
   c. The expectations of parent/spouse/worker become contradictory.
   d. Not having enough time or energy to do a task to the best of one's ability.
   e. Feeling stress because of wearing "too many hats."

__d__ 3. Which of the following is an example of an alternative work/family life cycle approach?
   a. The husband is considered a "helper" with household duties.
   b. The husband is regarded as the major economic provider with little responsibility for family work.
   c. The wife is primarily responsible for care-giving and household tasks.
   d. The family tries to reduce role-strain and overload by reallocating traditional family roles.
   e. all of the above

__d__ 4. The role of homemaker
   a. has been common in almost all societies since long before the industrial revolution.
   b. is associated with economic dependence.
   c. appears to contribute to the psychological well-being of middle-class women.
   d. is a relatively recent invention, which has come about since industrialization.
   e. all of the above

Marriage, Work, and Economics

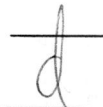

___d___ 5. In dual-earner marriages
   a. the husband's contribution to household work significantly increases if his spouse is employed.
   b. marital power is relatively unaffected by the wife's employment.
   c. both spouses are usually equally responsible for child care.
   d. husbands and wives generally agree that if both work outside the home, house hold tasks should be divided equally.
   e. both a and d

___e___ 6. Which of the following is/are true regarding women's employment patterns?
   a. The majority of women are employed in low-paying, low-status jobs in which women are overrepresented.
   b. Employed mothers generally do not seek personal fulfillment in their work as much as they do additional family income.
   c. Low-paying, low-status jobs offer women more flexibility for planning families than professional and managerial work.
   d. Women traditionally tend to interrupt their work lives significantly more than men.
   e. all of the above

___c___ 7. Which of the following is not included in the four possible sequential work/family patterns working women may follow?
   a. the conventional pattern
   b. the early interrupted pattern
   c. the middle interrupted pattern
   d. the late interrupted pattern
   e. the unstable pattern

___d___ 8. Which of the following is not true regarding family issues in the workplace?
   a. Employed women are less depressed and anxious than nonemployed homemakers.
   b. Women are particularly vulnerable to sexual harassment.
   c. Occupational differences exist along ethnic lines which affect the economic well-being of families.
   d. Most corporations provide evidence that they esteem parenting.
   e. All of the above are true.

___d___ 9. Which of the following ideals/beliefs regarding family policy would be most likely to be supported by political liberals?
   a. Father knows best while mother bakes cookies.
   b. Welfare leads to family breakup.
   c. The family is the last bastion of privacy and should remain inviolable.
   d. The state has an obligation to protect and assist families.
   e. None of the above are liberal orientations regarding family policy.

Chapter 11

___C___ 10. Which of the following is true regarding poverty and welfare?
   a. Most welfare recipients are permanently on welfare.
   b. Poverty and unemployment are basically economic issues.
   c. About 25 percent of the U.S. population requires welfare assistance at some time in their lives
   d. Poverty has been steadily declining since 1981.
   e. Most children on welfare receive welfare when they leave home.

___e___ 11. Child care is a problem for working families because
   a. it may involve constantly switching arrangements.
   b. finding adequate care can be a difficult and frustrating task.
   c. many mothers are left without after-school child care and often worry about their latch-key children.
   d. child care responsibilities make it difficult to compete with men for equal employment.
   e. all of the above

___b___ 12. Which of the following is not true regarding the hungry and homeless in the United States?
   a. Today more then 25 percent of the homeless are families.
   b. The majority of women become homeless because they are evicted.
   c. Many shelters permit only women and children.
   d. Shelters are often frightening environments for families.
   e. All of the above are true.

PART II - True/False

___T___ 1. Even with sex roles changing so greatly, a woman's work molds itself to her family, but a man's family molds itself to his work.

___F___ 2. Work spillover mainly affects men.

___T___ 3. In the sequential work/family pattern, the woman juggles child-rearing and career back and forth over time.

___F___ 4. The majority of women between the ages of 24-35 do not work, but stay at home with their children.

___T___ 5. A wife who holds outside employment generally has more influence in the decision-making process than a wife who does not work outside of the home.

Marriage, Work, and Economics

___F___ 6. Wives who return to work generally find that husbands and children willingly increase their share of the household tasks in appreciation of the increased income.

___T___ 7. Women's economic self-sufficiency leads to a greater likelihood that they will leave an unsatisfactory marriage.

___T___ 8. About half of America's children are vulnerable to poverty spells at least once during their childhood.

___T___ 9. Businesses are usually run as if every worker were male with a wife at home to attend to his and the children's needs.

___T___ 10. Because of the great difference in women's and men's wages, many women are condemned to poverty and are forced to accept welfare and its accompanying stigma.

___F___ 11. Sex is often the dominant element of sexual harassment.

___T___ 12. Self-care of children exists in families of all socioeconomic classes.

___T___ 13. Role overload and interrole conflict are likely to be two problems facing the working parent.

___F___ 14. Poverty can only touch those who are on the lowest rungs of the social ladder.

___F___ 15. Most women who receive welfare are lifetime recipients and their children grow up to be recipients as well.

___F___ 16. Most of those who accept government assistance do so for five years or more.

___T___ 17. In terms of policy, families are not a national priority.

___F___ 18. The majority of women become homeless because they lose their jobs.

Chapter 11

DISCUSS BRIEFLY

1. How has poverty become an increasing issue for American families today?

2. How are employment, gender and ethnicity related to poverty.

3. What are some of the factors related to being single and being a mother that lead to the "feminization of poverty"?

4. Some marriage counselors feel that the conflict over housework is becoming a major issue for two-career families. Why is this becoming such an important issue?

Chapter 11

SELF-DISCOVERY

Do you plan on having (or do you now have) a career or job that may conflict with family responsibilities? What conflicts do you project (or currently face)?

How is (would) your life be affected by role strain? What ways do you feel you can best cope with your situation?

How does (would) your gender affect your job situation?

JUST FOR FUN

The following survey examines your attitudes toward gender roles and the dual-career family. Please mark each statement with one of the following responses:

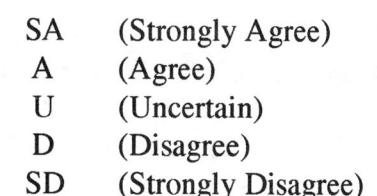

        SA   (Strongly Agree)
        A   (Agree)
        U   (Uncertain)
        D   (Disagree)
        SD   (Strongly Disagree)

There is no "right" or "wrong" answer, only your opinion. At the end of this chapter, there is a key to enable you to determine if you tend to hold traditional views, equalitarian views or some combination of the two.

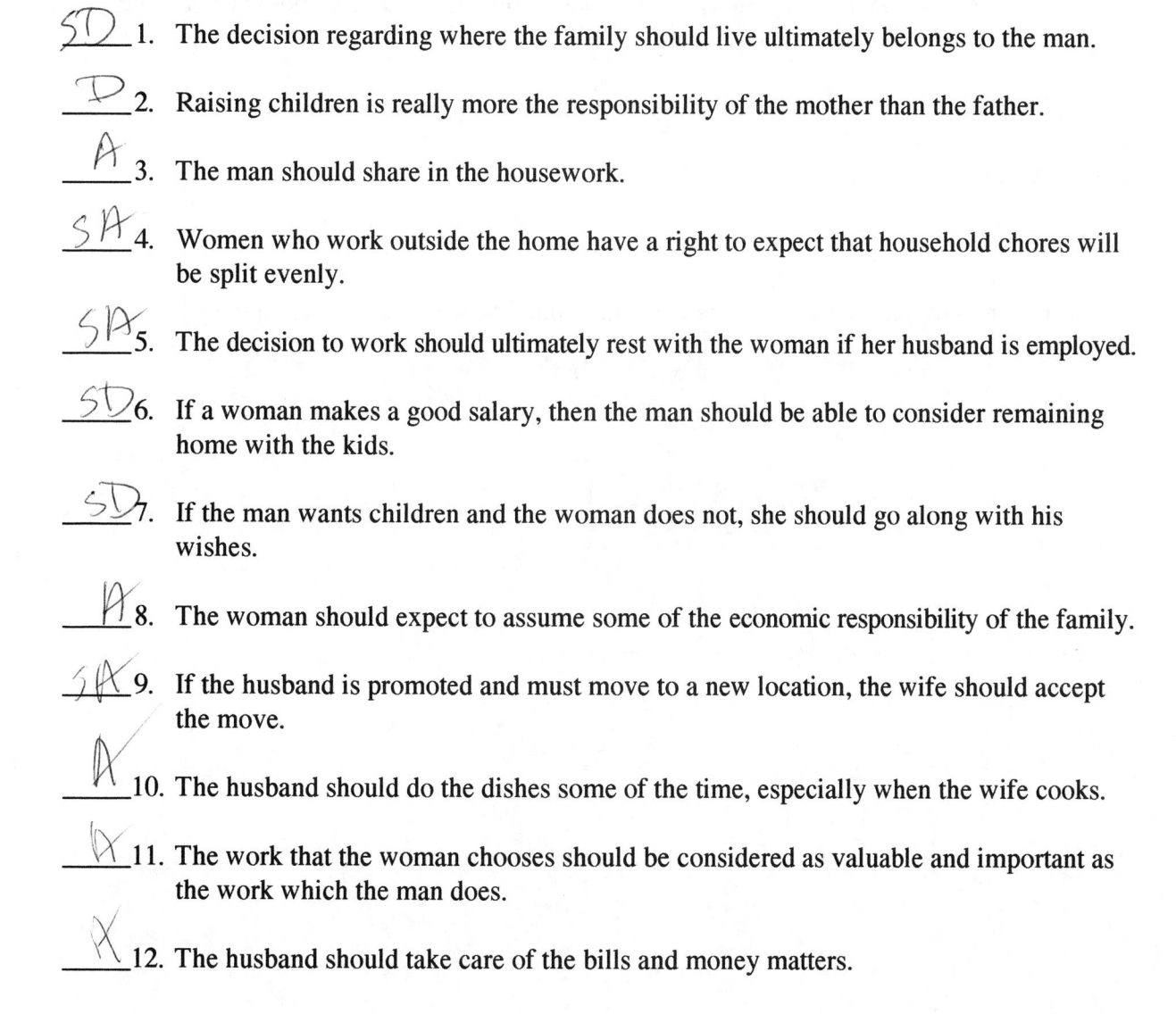

_SD_ 1. The decision regarding where the family should live ultimately belongs to the man.

_D_ 2. Raising children is really more the responsibility of the mother than the father.

_A_ 3. The man should share in the housework.

_SA_ 4. Women who work outside the home have a right to expect that household chores will be split evenly.

_SA_ 5. The decision to work should ultimately rest with the woman if her husband is employed.

_SD_ 6. If a woman makes a good salary, then the man should be able to consider remaining home with the kids.

_SD_ 7. If the man wants children and the woman does not, she should go along with his wishes.

_A_ 8. The woman should expect to assume some of the economic responsibility of the family.

_SA_ 9. If the husband is promoted and must move to a new location, the wife should accept the move.

_A_ 10. The husband should do the dishes some of the time, especially when the wife cooks.

_A_ 11. The work that the woman chooses should be considered as valuable and important as the work which the man does.

_A_ 12. The husband should take care of the bills and money matters.

Chapter 11

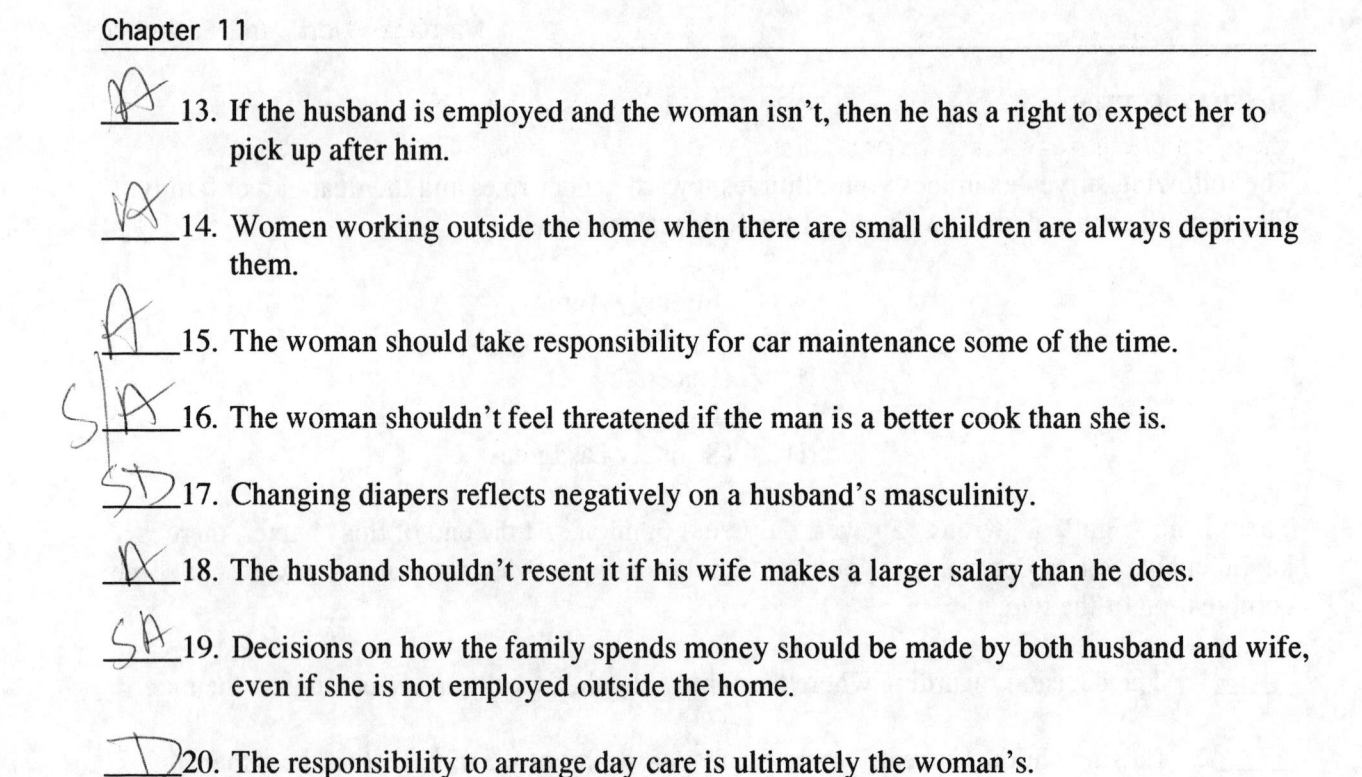

___A___ 13. If the husband is employed and the woman isn't, then he has a right to expect her to pick up after him.

___A___ 14. Women working outside the home when there are small children are always depriving them.

___A___ 15. The woman should take responsibility for car maintenance some of the time.

___SA___ 16. The woman shouldn't feel threatened if the man is a better cook than she is.

___SD___ 17. Changing diapers reflects negatively on a husband's masculinity.

___A___ 18. The husband shouldn't resent it if his wife makes a larger salary than he does.

___SA___ 19. Decisions on how the family spends money should be made by both husband and wife, even if she is not employed outside the home.

___D___ 20. The responsibility to arrange day care is ultimately the woman's.

MINI-ASSIGNMENT

Examine your attitudes about family policy. Are your attitudes more conservatively or more liberally oriented?

Marriage, Work, and Economics

What kinds of family policies would be in agreement with your ideals? What impact would these policies have on America's families?

KEY TO SELF QUIZZES

Multiple Choice   **True/False**

1. d   1. T   13. T
2. c   2. F   14. F
3. d   3. T   15. F
4. d   4. F   16. F
5. d   5. T   17. T
6. e   6. F   18. F
7. c   7. T
8. d   8. T
9. d   9. T
10. c   10. T
11. e   11. F
12. b   12. T

Chapter 11

Scoring the JUST FOR FUN Section

Give yourself a point for each of your answers which agrees with the answers below.

NOTE: There is no "right" or "wrong" response; your response is "right" for you. The higher the score, the more liberal you are in your attitudes. Lower scores reflect a more conservative, traditional perspective. The most important thing is that both partners in a relationship share similar values.

1. SD, D
2. SD, D
3. SA, A
4. SA, A
5. SA, A, U
6. SA, A, U
7. SD, D
8. SA, A
9. SD, D, U
10. SA, A
11. SA, A
12. SD, D, U
13. SD, D
14. SD, D, U
15. SA, A
16. SA, A
17. SD, D
18. SA, A
19. SA, A
20. SD, D

SUGGESTED READINGS

For related readings see pages 406 - 407 in your text.

CHAPTER 12

Families and Wellness

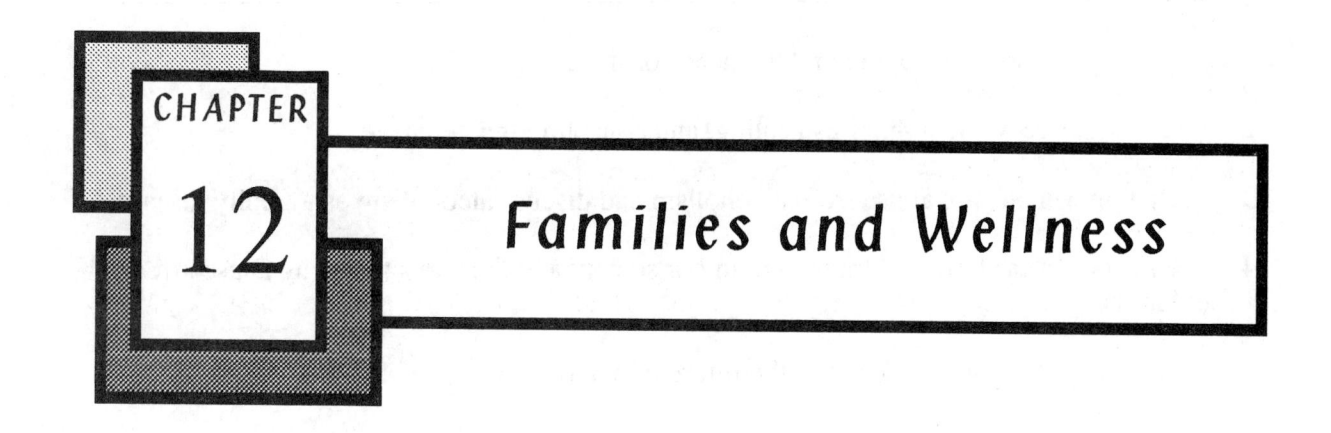

MAIN FOCUS

Chapter Twelve examines families and wellness; health care in crisis; stress and the family; family caregiving; death and dying in the United States; alcohol and drug abuse in the family; and the future of health care.

GOALS OF THIS CHAPTER

To demonstrate mastery of this chapter, you should be able to:

1. Explain how the family context influences wellness.

2. Explain how marriage may enhance mental and physical health.

3. Describe the crisis in American health care.

4. Explain the influence of gender, ethnicity, and socioeconomic status on wellness.

5. Discuss the connection between health insurance and wellness.

6. Define stress and the three types of stress.

7. Identify the four types of stressors which can provoke family stress.

8. Describe and explain the Double ABC-X model of family stress.

9. Explain stresses experienced by families caring for ill or disabled members and suggested coping strategies.

10. Explain the challenges involved in family caregiving and elder care.

Chapter 12

11. Identify the stages of dying and the needs of the dying.

12. Discuss bereavement, grief as healing, and consoling the bereaved.

13. Distinguish alcohol abuse from alcoholism and discuss alcoholism as a family disease.

14. Understand the types of drugs used in our society and the ways that drug abuse affects the family.

15. Explain policy issues related to the future of health care.

KEY TERMS AND IDEAS

The following terms, ideas, and concepts are listed in the order in which they appear in Chapter Twelve and in the outline. Be sure that you understand and can define each of the following:

stress	double ABC-X model	bereavement
eustress	stressor pileup	shiva
neustress	caregiver role	alcohol abuse
distress	thanatologist	alcohol dependence
stressor	hospice	alcoholism

CHAPTER TWELVE OUTLINE

I. INTRODUCTION TO THE CHAPTER
   A. Families consist of individuals with bodies: Our physical health and emotional health are critical elements of family life.
   B. During the last twenty years, research on the interrelationship of families and health has burgeoned.
   C. The family affects an individual's health and vice versa.

II. FAMILIES AND WELLNESS
   A. In a healthy family, the six dimensions of wellness interrelate to create and support optimal health and well-being.
   B. The relationship between the family and health should be viewed as interpenetrating rather than causal.
   C. It is within our families that we receive our earliest and most powerful messages regarding healthy behaviors and risk reduction.
      1. Half of all premature deaths are behavior related and could be prevented.
      2. The family's emotional climate affects the health of its members.
      3. Declining family health appears to affect marital quality negatively in several ways.

D. Married men and women tend to be healthier and happier than their unmarried peers; they live longer, are less depressed, and have a higher general sense of well-being.
   1. Fu and Goldman found that unhealthy behaviors and characteristics influence marriage rates.
   2. Researchers suggest that marriage encourages people to be healthy by enabling them to participate in certain behaviors or situations that promote good health: (1) living with a partner; (2) social support; and (3) economic well-being.

III. HEALTH CARE IN CRISIS
   A. Because the United States does not offer universal health-care coverage, the greatest determinant of health is socioeconomic status: Many insured Americans don't seek treatment because of high deductibles and copayments.
   B. Our ethnicity or race per se does not determine our health as much as does our economic status.
      1. The factors that affect health the most are socioeconomic: Poorer people have less access to preventative care and treatment.
      2. Many victims of our inadequate health-care system are children.
      3. Poor children are at increased risk for HIV infection.
   C. Racial and ethnic minorities suffer disproportionate exposure to hazardous substances.
   D. Health insurance has become one of the prime determinants of an individual's or family's access to medical care.
      1. In 1994, 40 million Americans were uninsured at any given time.
      2. Health insurance is closely tied to the workplace.
      3. Many employees experience "job lock" and need to remain with current employers for fear of loosing health insurance.
      4. Politically, the debate around health care centers on (1) whether all Americans should be covered by health insurance, and (2) how the health insurance is to be paid.
      5. The effects of the insurance crisis are: (1) to force the uninsured to live in constant dread of a major medical expense; (2) to lead to deteriorating health; and (3) to affect the ability of our public hospitals and emergency rooms to function.

IV. STRESS AND THE FAMILY
   A. Life involves constant change which brings stress: Because all families undergo periods of stress, the ability to cope successfully is seen as an important factor in measuring family health.
   B. Families constantly face stress which is tension resulting from real or perceived demands that require the family system to adjust or adapt its behavior.
      1. Not all stress is bad for you: Many believe that humans need some degree of stress to stay well.
      2. Stress can be beneficial when it is a positive motivator.
      3. There are three types of stress: **eustress** (good stress), **neustress** (stimuli that have no consequences), and **distress** (bad stress).

Chapter 12

   C. There are six variables affecting a family's response to stress: **stressors**, family hardship, strains, resources, meaning, and coping.
   D. Families that are unable to cope find themselves moving from a state of stress into crisis. Once in crisis, the family system becomes immobilized and the family can no longer perform its functions.
   E. Stressors, events which provoke stress, may be (1) normative or nonnormative, (2) external or internal, (3) short term or long term, and (4) with norms or normless.
   F. The **double ABC-X model** of family stress is the most widely used model to explain family stress.
      1. A key concept in this model is the **stressor pile up**: During stress or crisis the family responds not only to a current stressor, but also to family hardships and strains.
      2. Stressor pileup magnifies the impact of the stressor event, transforming a relatively minor stressor into a catastrophic event.
      3. The double ABC-X model has the following components: "A" represents stressor pileup; "B" represents the family's coping resources; "C" represents the family's perception of the stressor pileup; and "X" represents the outcome of the situation.

V. FAMILY CAREGIVING
   A. One of the principal tasks of families is to care for their members in times of ill health or incapacity.
   B. Chronic illnesses, physical limitations/disabilities, and developmental impairments are ongoing: Care is usually given at home by the family.
   C. The family must shape itself around the needs, limits, and potentials of the ill or disabled family member.
   D. General stresses experienced by families caring for ill or disabled members include:
      1. strained family relationships,
      2. modifications in family activities and goals,
      3. increased tasks and time commitments,
      4. increased financial costs,
      5. special housing requirements,
      6. social isolation,
      7. medical concerns, and
      8. grieving over disabilities, limitations, and restricted life opportunities.
   E. Strategies for coping with serious health problems in the family include: (1) making a place for the illness and keeping the illness in its place; (2) keeping communication open; (3) developing good working relationships with health-care professionals; and (4) cultivating sources of support.
   F. Most studies suggest that chronic illnesses have negative effects on the family; others find positive or inconsequential effects.
   G. Most elder care giving is provided by women, generally daughters or daughters-in-law.
      1. Most adult children in a given family participate in parental caregiving in some fashion.
      2. Excessive support received from adult children may be harmful: It is important to define the level of care which is both appropriate and necessary.

H. Even though elder care is often done with love, it can be the source of profound stress: Caregivers often experience conflicting feelings about caring for an elderly adult relative.
I. Affection eases the burdens, but does not necessarily decrease the strains of caregiving that relatives experience.
   1. Caregiver education and training programs, self-help groups, caregiver services, and family therapy can provide assistance in dealing with problems encountered by caregivers.
   2. Because elder care involves complex emotions raised by issues of dependency, adult children and their parents often postpone discussions until a crisis occurs.
   3. The best way for adult children to deal with elder care is to plan ahead with their siblings and aging parents, if they are willing.

VI. DEATH AND DYING IN AMERICA
   A. Even though we cognitively know that death comes to us all, when we actually confront death or dying, we are likely to be surprised, shocked, or at a loss about what to do.
   B. Our cultural and religious context is important in determining our response to death.
   C. American responses to death and dying fall into three categories: denial, exploitation, and romanticization.
   D. Many of us share a number of fears and anxieties regarding the dying process and death itself.
   E. **Thanatologists** (those who study death and dying) tell us that we need to develop a realistic, honest view of death as part of life: Acknowledging that death exists can greatly enrich our lives.
   F. The stages of dying are: denial and isolation, anger, bargaining, depression, and acceptance.
   G. Aside from basic physical care and relief from pain, a dying person needs most of all to be treated as a human being, yet death and dying in the hospital setting is generally characterized by impersonality.
   H. The hospice movement tries to counteract this impersonality. A **hospice** is a medical program that emphasizes both patient care (including management of pain and symptoms) and family support.
   I. **Bereavement** is our response to the death of a loved one.
      1. Our culture, religion, and personal beliefs all influence the type of mourning rituals we participate in after someone dies.
      2. Under Jewish law, there are three successive periods of mourning: **shiva**, shloshim, and avelut.
      3. Among Latinos, el Dia de los Muertos (the Day of the Dead), an ancient ritual to honor the dead, is making a comeback.
   H. Grieving is a process which may include shock, denial, depression, anger, loneliness, and feelings of relief: Guilt is often experienced as part of the grieving process.
      1. Healing, which is the goal of grieving, comes little by little as we work through grief.
      2. Being able to share one's grief with others is a crucial part of healing, thus, consoling the bereaved often involves "just" being there and, especially, listening.

Chapter 12

VII. ALCOHOL, DRUG ABUSE, AND FAMILIES
  A. Alcohol and drug abuse are the most prevalent and costly mental disorders currently facing society.
  B. Alcohol is a drug, however, many people do not think of alcohol as a drug.
    1. The abuse of alcohol can be seen as a symptom of a disorder that is both physical and emotional; alcoholism is a disease, not a personal shortcoming.
    2. **Alcohol abuse** involves continued use of alcohol despite awareness of social, occupational, psychological, or physical problems related to drinking or drinking in dangerous ways or situations.
    3. **Alcohol dependence**, or **alcoholism**, is a separate disorder involving more extensive problems than alcohol use.
    4. Although physiological and genetic factors may predispose a person to alcoholism, it appears that alcoholism is a learned behavior.
  C. Alcohol is the number one drug problem among our nation's youth (see page 440 of the text for other statistics).
    1. Millions of American families are affected by alcoholism, undermining their relationships and hopes.
    2. The popularly perceived image of the alcoholic as a "bum in the gutter" prevails, because families tend to deny their alcoholism.
  D. Alcoholism is sometimes called a "family disease" because it involves all members of the family in a "complex interactional system."
    1. The principle of homeostasis — the tendency toward stability in a system — operates in alcoholic families, maintaining established behavior patterns and strengthening resistance to change.
    2. Recovery is possible, but the alcoholic must face his or her denial (and often family denial) of the alcoholism, make a conscious choice to become well and to then seek treatment.
    3. For many families it is imperative that they be treated as a unit, because the family structure and stability may be organized around the alcoholism.
  E. There are many kinds of psychoactive drugs and many reasons for using them.
    1. Among the psychoactive drugs, the most commonly used include marijuana, hallucinogens, cocaine, PCP, narcotics, and inhalants.
    2. Psychotherapeutic drugs such as antidepressants, stimulants and sedatives are widely used.
    3. Drug use is prevalent in virtually all socioeconomic, sociocultural, and age groups (beginning with preteens).
    4. Drug use is a problem that must be dealt with on multiple levels: societal, familial, and individual.
  F. Researchers recognize the role of the family in beginning, continuing, stopping, and preventing drug use by its members.
    1. Some researchers believe that parents with marital problems use their adolescent's drug abuse as a means of avoiding their own problems: In such cases, drug abuse is merely a symptom of family pathology.

2. In families of chronic drug abuse, members are likely to feel anxious, depressed, guilty, or enraged: Communication is probably poor with little direct expression of emotion.
G. Drug abuse often begins as an adolescent problem.
1. The use of certain drugs by adolescents may be an important part of their peer's culture, making it especially difficult to "just say no."
2. In preventing and treating drug use, it is crucial to understand the social and cultural influences involved.
H. The family context is of paramount importance in the treatment of drug abuse.
1. Many drug abuse programs involve intensive family therapy.
2. Family members may sabotage the treatment of chronic drug users.
3. In addition to cultural factors, individual factors influence a person's drug use patterns and recovery process.
4. A person who understands his or her drug use from physiological, psychological and social perspectives is better equipped to combat it.

VIII. THE FUTURE OF HEALTH CARE
A. There remains much to be done to make adequate health care uniformly available to all Americans.
B. Policy issues that need to be addressed include (1) access to care, (2) poverty as a health issue, (3) preventative care and health maintenance, (4) mental health, (5) diversity, (6) support for families, and (7) united effort.

IX. READINGS AND FEATURES
A. In *Creating a Family Health Tree*, the authors illustrate how to compile and interpret a family health tree.
B. In *Women's Health: A Closer Look*, the authors discuss the gender biases, lack of information, and barriers women face in obtaining health care.
1. Women's traditional roles are related to women's health.
2. Research found that the likelihood of women receiving preventive services was influenced by insurance coverage, regular sources of care, financial barriers, and age.
3. Ethnicity appears to be an important factor in health-care access.
4. Violence is a major health issue for women.
C. In *Understanding Yourself: Do You Know Your Medical Rights*, the authors present select situations and explain medical rights related to those situations.
D. In *Understanding Yourself: Stress: How much Can Affect Your Health?*, the Social Readjustment Rating Scale is presented.
E. The *Perspective: HIV, AIDS and the Family* discusses how both the person with AIDS and his or her family experience social stigma and isolation.
1. They must cope with fear of contagion, infection and abandonment; feelings of guilt, anger and grief; and economic hardship.
2. Supporting persons with AIDS includes: showing caring by visiting and keeping in contact; calling before visiting; touching them; being willing to talk about the disease and prognosis; taking them out to dinner; offering assistance with chores; organizing your own support group; and being available.

Chapter 12

TEST YOUR COMPREHENSION

The chart below and on the facing page lists various concepts found in Chapter Twelve in your text. Complete the chart by defining each concept and giving an example of how it influences families and/or relationships.

Chart 12 - Part 1

CONCEPTS, DEFINITIONS, AND EXAMPLES FROM CHAPTER 12		
Term	Definition	Influence on Family or Relationship
dimensions of wellness		
social support		
environmental equality		
stress		
eustress		
neustress		

200

Families and Wellness

Chart 12 - Part 2

CONCEPTS, DEFINITIONS, AND EXAMPLES FROM CHAPTER 12		
Term	**Definition**	**Influence on Family or Relationship**
distress		
double ABC-X model		
stressor pileup		
caregiver role		
hospice		
bereavement		
alcohol abuse		
alcoholism		

Chapter 12

SELF QUIZZES

How well do you know this material? Test your understanding of the reading assignments by answering the following sample questions.

PART I - Multiple Choice: Choose the most correct response.

___C___ 1. All but which one of the following are true regarding the family context of health and illness?
   a. The family's emotional climate affects the health of its members.
   b. Our ethnicity affects our relationship to health and illness.
   c. The family is a secondary social agent in the promotion of health and well-being.
   d. The family has a powerful influence on health beliefs and behaviors.
   e. All of the above are true.

___d___ 2. Which of the following is not a way in which marriage encourages health and well-being?
   a. Married people are somewhat less likely to experience high levels of stress occasioned by the daily grind of poverty.
   b. A partner provides secondary prevention by helping to identify or treat an illness or disease early.
   c. Intimacy aids in a partner's recovery.
   d. Major health differences result from the presence of any another person (e.g. roommate, friend or spouse).
   e. Social support encourages and reinforces protective health behaviors.

___d___ 3. Which of the following is not true regarding health care in the United States?
   a. The greatest determinant of health is socioeconomic status.
   b. Many of the victims of our inadequate health-care system are children.
   c. Health insurance is a prime determinant of access to medical care.
   d. Our vast outlay of money for medical technology has lowered our infant mortality rate below most industrialized nations.
   e. The health status of many of our African-American communities is that of third world countries.

___a___ 4. In terms of families and wellness, the fact that factories and facilities that produce toxic wastes are concentrated in minority communities raises concerns about
   a. environmental equity.
   b. quality air control.
   c. environmental discrimination.
   d. environmental protection agencies.
   e. toxic waste dumps.

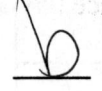

___b___ 5. The type of stress producing stimuli that have no consequential effects is called ___.
   a. quasi-stress
   b. neustress
   c. nilstress
   d. eustress
   e. pseudostress

202

Families and Wellness

___e___ 6. Being a lesbian parent could be an example of a _____ stressor.
   a. nonnormative
   b. external
   c. normless
   d. short term
   e. all but d

___d___ 7. According to the double ABC-X model, which of the following is not necessarily related to how a family experiences a crisis?
   a. the family's perception of the stressful event
   b. stressor pileup
   c. possession of resources
   d. distribution of power
   e. lack of economic resources

___c___ 8. Which of the following is not true about caregiving?
   a. Caregivers often experience conflicting feelings about caring for an elderly relative.
   b. Elder caregiving affects husbands and wives differently.
   c. Self-help groups appear to be the most effective for dealing with the emotional aspects of caregiving.
   d. Adult children and their parents often postpone discussions about dependency until a crisis occurs.
   e. Elders are especially vulnerable to abuse by their children.

___a___ 9. One of the greatest fears aroused by the thought of death is the fear of
   a. loosing control.
   b. separation.
   c. an end to pleasure.
   d. the unknown.
   e. isolation.

___b___ 10. Which of the following is not true regarding the abuse of alcohol?
   a. Alcohol use by pregnant women can result in fetal damage and fetal alcohol syndrome.
   b. Alcohol abuse is less of a problem than drug abuse.
   c. Over fifteen percent of adults meet criteria for alcohol abuse.
   d. Health professionals view alcoholism as a disease rather than a personal shortcoming.
   e. Alcohol is a factor in nearly half of all U.S. murders, suicides, and accidental deaths.

___b___ 11. The principle of _____ operates in alcoholic families to strengthen resistance to change and maintain behavior patterns.
   a. solidarity
   b. homeostasis
   c. denial
   d. loyalty
   e. entrenchment

Chapter 12 — HIV

___e___ 12. Treatment for alcoholism is most successful if
   a. the alcoholic makes the conscious choice to become well.
   b. the family supports the decision.
   c. the whole family participates in the treatment process.
   d. the alcoholic deals with enlightened professionals.
   e. All of the above are very important.

___d___ 13. Which of the following is not true regarding the family's role in drug abuse?
   a. Family members may sabotage the treatment of addicted members.
   b. Drug abuse may be a symptom of family pathology.
   c. Drug abusers remain intimately involved with their families.
   d. Parents often feel the need to take responsibility for their child's substance abuse.
   e. All of the above are true.

___e___ 14. According to the Commonwealth Fund Survey, the likelihood of women receiving preventative services was influenced by
   a. traditional gender roles.
   b. regular sources of care.
   c. age.
   d. both a and b
   e. all of the above

___e___ 15. Which of the following is a way to provide needed support for friends or relatives with HIV or AIDS?
   a. touching them
   b. ~~dropping in on them to visit~~
   c. finding an HIV or AIDS support group for them
   d. offering to talk to them about HIV/AIDS and its prognosis
   e. all but b

PART II - True/False

___T___ 1. Our physical and emotional health are critical elements of family life.

___F___ 2. The relationship between the family and health should be viewed as causal.

___T___ 3. Being happily married is good for one's health.

___T___ 4. Persons with unhealthy behaviors face a greater difficulty in finding an acceptable spouse.

___T___ 5. Poverty is associated with decreased life expectancy.

Families and Wellness

F   6. The greatest determinant of health is ethnicity.

T   7. Health insurance is one of the prime determinants of a family's access to medical care.

F   8. The ability to adapt to stress has little impact on family strength.

F   9. Dealing with a family member's disability rarely strengthens families.

F   10. Adult children can never do too much in terms of the social support they offer their elders.

T   11. Thanatologists believe that a certain amount of fear of death is healthy.

T   12. When consoling a bereaved person, it is helpful to talk about the person who died.

F   13. There is a lot less alcohol in a glass of wine than in a mixed drink.

T   14. Dysfunctional families often contribute to the development of alcoholism and drug addiction.

T   15. Drug use is prevalent in virtually all socioeconomic, sociocultural, and age groups preteen and over.

F   16. Chronic drug abusers are usually disengaged from their families.

DISCUSS BRIEFLY

1. What is the American health care crisis and how does it impact America's families?

Chapter 12

2. What are the different types of stressors and how do they impact families?

3. How do gender, ethnicity, and socioeconomic status influence wellness?

4. What are some conflicts experienced by primary caregivers and their families?

5. What are the critical issues regarding the future of health care?

SELF-DISCOVERY

Examine some of your attitudes and feelings about death and dying by answering the following questions:

1. What are some ways your culture, religion, and personal beliefs influence your mourning rituals? Your family's mourning rituals?

2. How can an understanding of the process of dying be beneficial to the dying person as well as their family members? The process of grieving?

Chapter 12

3. What are some problems experienced by people with AIDS and their families? Which of these would you find to be the most difficult?

4. What are some ways to console the bereaved? Which of these are you most comfortable with?

KEY TO SELF QUIZZES

Multiple Choice

1. c   10. b
2. d   11. b
3. d   12. e
4. a   13. d
5. b   14. e
6. e   15. e
7. d
8. c
9. a

True/False

1. T   10. F
2. F   11. T
3. T   12. T
4. T   13. F
5. T   14. T
6. F   15. T
7. T   16. F
8. F
9. F

SUGGESTED READINGS

For related readings, see pages 448 - 449 in the text.

CHAPTER 13
Family Violence and Sexual Abuse

MAIN FOCUS

Chapter Thirteen examines models of family violence; battering; the cycle of violence; marital rape; violence in dating and relationships; why women stay in violent relationships; interventions; and the hidden victims of family violence. It also discusses child abuse and neglect, child sexual abuse, treatment programs for sexually abused children, and preventing sexual abuse.

GOALS OF THIS CHAPTER

To demonstrate mastery of this chapter, you should be able to:

1. Describe the concepts of violence and abuse, and understand why family research into this topic has been so problematic.

2. Describe the cultural roots of violence in our society and how this may contribute to violence in the family.

3. Explain the principal models used in understanding family violence.

4. Define battering, the type of people likely to batter, the type of people likely to become victims of battering, and the conditions likely to contribute to the abuse.

5. Describe the cycle of violence which often characterizes the battering of a spouse.

6. Discuss marital rape and the difficulties in establishing this as a legal issue.

7. Explain how dating violence and acquaintance rape are problems in our society and what can be done to reduce risk to women.

Chapter 13

8. Explain why battered women stay in battering relationships.

9. Understand interventions and programs necessary to end domestic violence and cite the problems facing programs intended to reduce such violence.

10. Define child abuse and neglect and describe those families most likely to be at risk.

11. Discuss the problem of violence between siblings, violence against parents, and the abuse of the elderly.

12. Describe the general preconditions for sexual abuse.

13. Describe and define the forms of sexual abuse in families.

14. Explain the effects of sexual abuse, both initially and long term.

15. Discuss sexual abuse prevention programs.

KEY TERMS AND IDEAS

The following terms, ideas, and concepts are listed in the order in which they appear in Chapter Thirteen and in the outline. Be sure that you understand and can define each of the following:

violence	date rape	intrafamilial sexual abuse
battering	acquaintance rape	pedophilia
rape	child sexual abuse	incest
marital rape	extrafamilial sexual abuse	

CHAPTER THIRTEEN OUTLINE

I. INTRODUCTION TO THE CHAPTER
   A. Intimacy or relatedness in any form can increase the likelihood of violence or sexual abuse.
      1. Only war zones and urban riot scenes are more dangerous places than families.
      2. Every 30 seconds a woman is beaten by her boyfriend or husband.
      3. At least a million American children are physically abused or neglected each year.
   B. Even though domestic violence is beginning to be recognized and understood, there is much work to be done toward reducing and eliminating it.

II. FAMILY VIOLENCE
   A. Researchers have not been in complete agreement about what constitutes violence.
      1. The text uses the following definition of **violence** — "an act carried out with the intention or perceived intention of causing physical pain or injury to another person."
      2. Violence may be seen as a continuum, with "normal abuse" (e.g. spanking) at one end and abusive violence (acts with high potential for causing injury) at the other extreme.
   B. To better understand the roots of violence within the family, we must look at its place in the larger sociocultural environment.
      1. Aggression is a trait that our society labels as generally desirable, especially for males.
      2. Cultural scripts such as getting ahead at work, asserting ourselves in relationships, and winning at sports are all culturally approved and reward aggression.
      3. The psychiatric model finds the source of family violence within the personality of the abuser and assumes that the individual is violent as a result of mental or emotional illness, psychopathology, or perhaps alcohol or drug misuse.
      4. The ecological model uses a systems model to look at the child's development within the family environment and the family's development within the community.
      5. The patriarchy (male dominance) model takes a historical perspective by holding that most social systems have traditionally placed women in a subordinate position to men, thus supporting male violence.
      6. The social situational and social learning models view the sources of violence as originating in the social structure.
         a. The social situational model views family violence as arising from structural stress and cultural norms.
         b. The social learning model holds that people learn to be violent from society at large and from their families.
      7. The resource model assumes that social systems are based on force or the threat of force.
      8. The exchange/social control model is based on a two-part theory: (1) exchange theory holds that in our interactions we constantly weigh the perceived rewards against the costs, and (2) social control raises the costs of violent behavior through such means as arrest, imprisonment, loss of status or loss of income.
   C. Battered Women and Battering Men
      1. **Battering** includes slapping, punching, knocking down, choking, kicking, hitting with objects, threatening with weapons, stabbing, shooting, and sexual abuse.
      2. Battering women is one of the most common and underreported crimes in the United States.
         a. About one woman in twenty-two is the victim of abusive violence each year.
         b. One-third of murdered women are killed by their husbands or lovers.
         c. Women of all races are equally vulnerable to attacks by intimates.
      3. Wife abuse is more common and more severe in families of lower socioeconomic status and relationships with a high degree of marital conflict.

Chapter 13

4. A man who systematically batters is likely to: have low self-esteem; believe the myths about battering and traditional sex roles; have a dual personality; be sadistic, passive-aggressive, or pathologically jealous; use sex as an act of aggression; or believe in the moral rightness of his violent behavior.
5. Although some men have been injured by wives, "husband battering" is probably a misleading term.

D. The three-phase cycle of violence involves the building of tension, the explosion, and the "honeymoon period" (where the batterer begs forgiveness and promises never to batter again).
   1. Alcohol is involved in the majority of cases.
   2. In a battering relationship, the woman suffers serious physical and emotional damage.

E. Research indicates that the rate of abuse in gay and lesbian relationships is comparable to that of heterosexual relationships, however, there is often nowhere to go for support.

F. **Marital rape** is one of the most widespread and overlooked forms of family violence, but many people (including the victims themselves) have difficulty acknowledging that forced sex in marriage is rape.
   1. There is a notion that the male breadwinner should be the beneficiary of some special immunity because of his family's dependence on him.
   2. Marital rape victims experience feelings of betrayal, anger, humiliation, and guilt.

G. Evidence suggests that dating violence approaches or even exceeds the level of marital violence: It appears that issues in dating violence are different than spousal violence.
   1. **Acquaintance rape (date rape)** is the most common form of rape: Alcohol or drugs are often involved.
   2. Physical violence often goes hand-in-hand with sexual aggression: Three-fourths of victims who were acquaintance rape victims sustained bruises, cuts, black eyes, and internal injuries.
   3. Much sexual communication is done nonverbally and ambiguously, creating considerable confusion and argument about sexual consent.

H. Reasons battered women stay in or return to violent situations include:
   1. economic dependence,
   2. religious pressure to submit to her husband's will,
   3. belief that the children need a father,
   4. fear of being alone, especially as she may have been cut off from all other ties,
   5. belief in the American dream of family bliss,
   6. pity for her husband,
   7. guilt and shame, feeling that it is somehow her own fault,
   8. duty and responsibility to stick it out,
   9. fear for her life if she tries to escape,
   10. love for the partner, despite the battering, and
   11. cultural reasons.

I. Learned helplessness is the result of behavioral reinforcement.
   1. It is connected to low self-esteem and keeps battered women feeling that they cannot control the battering
   2. The woman's determination that the violence must cease is crucial to stopping it.
J. Professionals have long debated the relative merits of compassion versus control as intervention strategies.
   1. Long ignored, domestic violence has only recently become a top concern for legislators and law enforcement agencies.
   2. At a point where a woman finds she can leave an abusive relationship, she may have serious needs such as: immediate medical attention and physical protection; accommodations for herself and/or children; access to support, counseling, assistance in the form of money, food stamps and other basic survival items; and a support system of professionals who are informed and compassionate.
      a. Shelters help women to network with other women who have been battered and provide many other services for battered women who call, such as information, advice, and referrals.
      b. Treatment services for men who batter provide one important component of a coordinated response to domestic violence
K. More than a million American children are physically abused by their parents each year.
   1. Child abuse was not recognized in the United States until the 1960s when Kempe and his colleagues coined the term battered-baby syndrome.
   2. Research indicates that nearly three out of four child slayings in the industrial world occur in the United States: Every five hours a child dies from abuse or neglect.
   3. Parental violence is among the five leading causes of death for children between the ages of one and eighteen.
   4. Research suggests that three sets of factors put families at risk for child abuse and neglect: parental characteristics, child characteristics, and the family ecosystem.
      a. Characteristics of parents who abuse their children may include: physical punishment by one's own parents; belief in corporal punishment; interspousal violence; belief that a father is the dominant authority figure; low self-esteem; unrealistic expectations of a child; persistent role reversal; and lack of concern about the seriousness of a child's injury.
      b. Children who are abused are often labeled by their parents as "unsatisfactory" and may be: a "normal" child (who is the product of a difficult or unplanned pregnancy or who is the wrong sex); an "abnormal" child (premature, possibly with congenital defects or illness); or a "difficult" child (e.g. who may show traits of fussiness or hyperactivity).
      c. The family ecosystem may include serious problems contributing to stress: unemployment; social isolation; low levels of income; unsafe and violent neighborhoods; a home that may be crowded, hazardous, dirty or unhealthy; an overburdened single-parent; or family health problems.

Chapter 13

5. Our system does not provide the human and financial resources necessary to deal with these socially destructive problems.
    a. With heightened public awareness in recent years and mandatory reporting of suspected child abuse in all fifty states, identifying abused children is much easier now than it was two decades ago.
    b. Much of the intervention in child abuse appears to be equivalent to putting a Band-Aid on a huge malignant tumor.
L. Considerable violence exists between siblings, teenage children and their parents, and adult children and aging parents.
    1. Sibling violence is by far the most common form of family violence and is often taken for granted in our culture.
        a. Variables affecting sibling violence include the increasing age of the child, the sex of the child, and the role models of their parents.
        b. It is possible to provide an environment in which violent alternatives are non-acceptable.
    2. Teenage violence towards parents has only recently been studied, but it may be as prevalent as spouse abuse.
    3. Of all the forms of hidden family violence, only the abuse of elderly parents by their grown children (or grandchildren) has received considerable public attention.
        a. Nevertheless, such abuse is often unnoticed, unrecognized, and unreported.
        b. The most likely victims of elder abuse are the very elderly who are suffering from physical and mental impairments.
    4. Prevention strategies for dealing with family violence usually focus on eliminating social stress or strengthening families including: reducing societal sources of stress; eliminating sexism; initiating prevention and early intervention; ending social isolation; breaking the family cycle of violence via education; and eliminating cultural norms that legitimize family violence.

III. CHILD SEXUAL ABUSE
   A. **Child sexual abuse** is defined as any sexual interaction (including fondling, erotic kissing, or oral sex) between an adult or adolescent and a prepubertal child.
   B. Child sexual abuse is generally categorized in terms of kin relationship; **extrafamilial sexual abuse** is perpetrated by nonrelated individuals while **intrafamilial sexual abuse** is perpetrated by related individuals, including steprelatives.
   C. **Pedophilia** is an intense, recurring sexual attraction to prepubescent children.
   D. According to David Finkelhor, there are four factors which are general preconditions for sexual abuse: (1) motivation to sexually abuse a child; (2) overcoming internal inhibitions against acting on the motivation; (3) overcoming external obstacles to committing sexual abuse; and (4) undermining or overcoming the child's potential resistance to the abuse.
   E. The **incest** taboo, which is nearly universal in human societies, prohibits sexual activities between closely related individuals.

F. There are different forms of intrafamilial sexual abuse including: father/daughter abuse (which is the most traumatic form and includes sexual abuse by a stepfather); brother/sister sexual abuse; and uncle/niece sexual abuse.
G. Children at risk are more likely to be female, preadolescent, have absent or unavailable parents, have poor relationships with parents, have parents who are in conflict, and children who live with a stepfather.
H. Numerous well-documented consequences of child sexual abuse exist for intrafamilial and extrafamilial abuse.
   1. Many abused children experience symptoms of post-traumatic stress disorder.
   2. The initial effects of sexual abuse include emotional disturbances, physical consequences, sexual disturbances, and social disturbances.
   3. The long-term effects of sexual abuse include depression, self-destructive tendencies, somatic disturbances, negative self-concept, interpersonal relationship difficulties, revictimization, and sexual problems.
I. Finkelhor and Browne suggest a model of sexual abuse trauma which includes traumatic sexualization, betrayal, powerlessness, and stigmatization.
J. Sexual abuse treatment programs work to treat the individual, the father/daughter relationship, the mother/daughter relationship, and the family as a whole.
K. The idea of preventing sexual abuse is relatively new, with prevention programs beginning about a decade ago.
L. Programs preventing sexual abuse have been hindered by the differing concepts of appropriate sexual behavior, the difficulties of discussing sexual abuse with children, and the controversy over sex education.
   1. Child abuse prevention (CAP) programs have aimed at teaching children that they have rights; including the right to control their own bodies and genitals, the right to feel "safe," and the right not to be touched in ways that feel confusing or wrong.
   2. Other CAP programs are directed at parents to help them educate their children.
   3. CAP programs for professionals encourage them to watch for signs of sexual abuse and to investigate children's reports of abuse.
M. In recent years, both the American Medical Association (AMA) and the federal government have become more actively involved in fighting domestic violence.

IV. READINGS AND FEATURES
A. In *The Mythology of Violence and Sexual Abuse*, the authors point out that the understanding of family violence and sexual abuse is often obscured by myths; twelve popular myths about family violence and sexual abuse in our society are exposed.
B. The *Perspective: The Epidemic of Missing Children: Myth or Reality?* suggests that most missing children are runaways, throw-aways, or abducted by parents in child custody disputes.

Chapter 13

TEST YOUR COMPREHENSION

The following chart examines the current models of family violence. For each model listed give the basic assumptions promoted by the model and the key factors related to the model.

Chart 13

MODELS OF FAMILY VIOLENCE		
Model	Basic Assumptions	Key Factors
Psychiatric		
Ecological		
Patriarchy		
Social Situational and Social Learning		
Resource		
Exchange/ Social Control		

Family Violence and Sexual Abuse

SELF QUIZZES

How well do you know this material? Test your understanding of the reading assignments by answering the following sample questions.

PART I - Multiple Choice: Choose the most correct response.

___e___ 1. Which of the following is true regarding family violence?
    a. Only war zones and urban riot scenes are more dangerous places than families.
    b. Every 30 seconds a woman is beaten by her boyfriend or husband.
    c. At least a million parents a year are physically assaulted by their adolescent children.
    d. Some boyfriends, husbands, and parents seem to believe that intimacy confers a "right" to be physically or sexually abusive.
    e. All of the above are true.

___b___ 2. Models used in understanding family violence are based on all but which one of the following concepts?
    a. A child who doesn't match well with the parents and a family that is under stress while having little community support is at increased risk for child abuse.
    b. Violent men lose status among their peers for asserting their authority.
    c. Social systems are based on force or the threat of force.
    d. Most social systems have traditionally placed women in a subordinate position to men.
    e. The source of family violence is within the personality of the abuser.

___d___ 3. Battering
    a. occurs mostly in families of low social status.
    b. is a relatively new phenomenon.
    c. is usually never endorsed by wives.
    d. is one of the most common and underreported crimes in our country.
    e. all but b

___a___ 4. Which of the following is not a characteristic of a man who batters?
    a. He hates and resents his wife.
    b. He has low self-esteem.
    c. He believes in traditional gender-role stereotypes.
    d. He believes in the moral rightness of his behavior.
    e. He may use sex as an act of aggression.

___c___ 5. According to Murray Straus, all but which one of the following are reasons for taking female violence seriously?
    a. Assaulting a spouse is morally wrong.
    b. Not doing so unintentionally validates cultural norms condoning violence.
    c. Female violence overshadows male violence.
    d. There is a danger of escalation.
    e. Spousal assault models violent behavior for children.

217

Chapter 13

___e___ 6. Marital rape
   a. is one of the most widespread and overlooked forms of family violence.
   b. is now considered a crime in most states.
   c. has not been regarded as a serious form of assault.
   d. may result in feelings of humiliation, betrayal and intense rage.
   e. all of the above

___b___ 7. Which of the following is characteristic of dating violence?
   a. Both men and women engage in violence, with greater injury inflicted on the men.
   b. Sexual violence may be more prevalent than physical violence.
   c. Counselors, physicians, and law enforcement agencies are commonly used by victims of dating violence.
   d. Dating violence occurs at a lower rate than marital violence.
   e. Premarital violence is a good indication of the depth of love in the relationship.

___d___ 8. To reduce the risk of date rape, women should consider all of the following except
   a. being forceful and firm.
   b. avoiding the use of alcohol.
   c. avoiding ambiguous verbal and nonverbal behavior.
   d. being polite.
   e. sharing expenses.

___b___ 9. Which of the following is not true of women who stay in violent relationships?
   a. They are often economically dependent on the batterer.
   b. They are masochistic and like being hurt.
   c. They believe they may be killed if they try to escape.
   d. They feel they have no safe place to go.
   e. They stay as a result of "learned helplessness," dependent on their partners.

___d___ 10. Concerning battered women and legal issues,
   a. ~~domestic violence has long been a top concern for legislatures.~~
   b. there is still resistance by some law enforcement and judicial branches to listen to the victims.
   c. professionals dealing with domestic violence have long debated the relative merits of control versus compassion as intervention strategies.
   d. both b and c
   e. ~~all of~~ the above

___a___ 11. Which of the following is not characteristic of families at risk for child abuse?
   a. The abusing father was physically punished by his parents.
   b. The parents believe in corporal discipline of children and wives.
   c. One or more family members have health problems
   d. The parents have a low educational level.
   e. The family lives in an unsafe neighborhood.

Family Violence and Sexual Abuse

__e__ 12. Sibling violence
   a. is by far the most common form of family violence.
   b. is taken for granted in our culture.
   c. decreases with the increasing age of the child.
   d. reflects their parents' actions.
   e. all of the above

__d__ 13. All of the following are true regarding the abuse of the elderly except
   a. it is the only form of hidden family violence which has received considerable public attention.
   b. it is estimated that approximately 500,000 elderly people are physically abused annually.
   c. much abuse of the elderly goes unnoticed, unrecognized and unreported.
   d. it is caused by the elderly person being an abusing parent in earlier times.
   e. the most likely victims are the very elderly who are suffering from physical or mental impairments.

__a__ 14. Regarding intrafamilial child sexual abuse, all of the following are true except
   a. there is an association between sexual abuse and socioeconomic status.
   b. the majority of sexually abused children are girls.
   c. the mother is especially important in protecting children.
   d. there is no association between sexual abuse and race.
   e. children with stepfathers are at greater risk for sexual abuse.

__c__ 15. Which of the following is not included in Finkelhor and Browne's "traumatic model of sexual abuse?"
   a. traumatic sexualization
   b. betrayal
   c. self-destructiveness
   d. powerlessness
   e. stigmatization

__c__ 16. All of the following are myths related to family violence and sexual abuse except
   a. family violence is restricted to families with low levels of education.
   b. violent spouses or parents have psychopathic personalities.
   c. most child sexual abuse is perpetrated by people the victim knows.
   d. abused children will grow up to abuse their own children.
   e. a battered woman can always leave home.

Chapter 13

PART II - True/False

_____ 1. Those nearest and dearest are also those most likely to be the victims of the violence-prone.

_____ 2. Aggression is a trait that is generally promoted in our culture as undesirable.

_____ 3. Over fifty percent of murdered women are killed by their husbands or lovers.

_____ 4. African-American couples experience more abuse than do white couples.

_____ 5. Factors such as self-esteem or childhood experiences of violence do not appear to be necessarily associated with a woman being in a battering relationship.

_____ 6. Socioeconomic status does not appear to be a significant factor in instances of family violence.

_____ 7. Husband battering is a significant problem that has not been significantly addressed.

_____ 8. Research indicates that rates of abuse are lower in gay and lesbian relationships than in heterosexual relationships.

_____ 9. Rape in marriage is one of the most widespread and overlooked forms of family violence.

_____ 10. Date rapes are usually planned.

_____ 11. "Acquaintance rape" is only a small part of all reported rapes.

_____ 12. Her belief in the American dream may keep a woman in a violent relationship.

_____ 13. Domestic violence has only recently become a top concern for legislators and law enforcement agencies throughout the country.

_____ 14. Much of the intervention in child abuse appears to be successful in containing and reducing child abuse.

_____ 15. Our social system does not currently provide the human and financial resources to deal with these socially destructive problems.

_____ 16. Violence between siblings is common in most American families.

_____ 17. Abuse of the elderly has received considerable attention, but often goes unnoticed and unreported.

_____ 18. The incest taboo is nearly universal in human societies.

_____ 19. The majority of missing children have been kidnapped by a stranger.

DISCUSS BRIEFLY

1. What factors appear significant in dating violence and rape? How can women protect themselves?

2. Why do women stay in a situation where they are battered?

2. How common is incest? What can we do as a society to deal with it?

Chapter 13

SELF-DISCOVERY

What are your feelings about corporal punishment for children? What are the authors' feelings? How do you draw the line between punishment and abuse?

MINI-ASSIGNMENT

Investigate a family service agency in the area in which you live which deals with one of the topics of family dysfunction discussed in this chapter. What is the agency? What services do they provide and to whom?

KEY TO SELF QUIZZES

Multiple Choice   **True/False**

Multiple Choice		True/False	
1. e	10. d	1. T	10. F
2. b	11. d	2. F	11. F
3. d	12. e	3. F	12. T
4. a	13. d	4. T	13. T
5. c	14. a	5. T	14. F
6. e	15. c	6. F	15. T
7. b	16. c	7. F	16. T
8. d		8. F	17. T
9. b		9. T	18. T
			19. F

SUGGESTED READINGS

For related readings, see pages 484 - 485 in the text.

CHAPTER 14

Coming Apart: Separation and Divorce

MAIN FOCUS

Chapter Fourteen examines factors affecting the likelihood of divorce; the process of divorce, marital separation; consequences of divorce; children and divorce; child custody; and divorce mediation.

GOALS OF THIS CHAPTER

To demonstrate mastery of this chapter, you should be able to:

1. Discuss trends in divorce and present divorce rates.

2. Discuss factors affecting the likelihood of divorce.

3. Explain the divorce process and Bohannon's six "stations" of divorce.

4. Discuss the process of uncoupling.

5. Describe the stages most individuals experience while undergoing divorce and the problems and emotions that they must face.

6. Discuss the functions of dating after divorce and how post-divorce dating is different from premarital dating.

7. Describe no-fault divorce, its advantages and disadvantages.

Chapter 14

8. Explain the economic consequences of divorce on women and children.

9. Discuss the psychological and social effects divorce may have upon children and how children may respond to the marital breakup of their parents.

10. Detail the major types of custody and the basis for awarding custody.

11. Describe the divorce mediation process.

12. Understand the relationship between gender and divorce-related stressors.

13. Describe the divorce experience of lesbians and gay men.

KEY TERMS AND IDEAS

The following terms, ideas, and concepts are listed in the order in which they appear in Chapter Fourteen. Be sure that you understand and can define each of the following:

social integration	child support	joint physical custody
separation distress	sole custody	split custody
no-fault divorce	joint custody	divorce mediation
alimony	joint legal custody	

CHAPTER FOURTEEN OUTLINE

I. INTRODUCTION TO THE CHAPTER
    A. Americans' feelings about marriage and divorce seem strangely paradoxical.
        1. Our high divorce rate may actually reflect an idealization of marriage.
        2. Divorce may be a critical part of contemporary marriages which emphasize fulfillment and satisfaction.
        3. Divorce is a persistent fact of the American marital and family life cycle, and one of the most important forces affecting and changing American lives and families today.
    B. In 1974, a watershed was reached when more marriages ended by divorce than by death.
        1. Today, approximately 50 percent of all new marriages are likely to end in divorce.
        2. About one in five American families is a single-parent family; more than half of all children will become stepchildren by the year 2000.
    C. Researchers traditionally looked on divorce from a deviance perspective: Social scientists are increasingly viewing divorce as one path in the normal family cycle.

Coming Apart: Separation and Divorce

II. FACTORS AFFECTING THE LIKELIHOOD OF DIVORCE
   A. It may be difficult to discover the underlying reasons for an individual divorce, but researchers have found various factors related to divorce.
   B. It is often difficult to view divorce in terms of societal factors because the pain of divorce seems so uniquely personal.
      1. The shift from an agricultural society to an industrial one undermined many of the family's traditional functions: As a result of losing many of its social and economic underpinnings, the family is not a necessity.
      2. **Social integration** — the degree of interaction between individuals and the larger community — is emerging as an important factor in the incidence of divorce: Geographic location in the U.S. is related to divorce rates.
      3. American culture has traditionally been individualistic: The individual is viewed by many as having priority over family when the two conflict.
   C. Demographic factors which appear to be related to divorce include employment status, income, educational level, ethnicity, and religion.
      1. Among whites, a higher divorce rate is characteristic of low-status occupations and unemployment.
      2. The higher the family income, the lower the divorce rate for both whites and African-Americans: The higher a woman's individual income, the greater her chances of divorce.
      3. For whites, the higher the educational level, the lower the divorce rate.
      4. African-Americans are more likely than whites to divorce: This may be due to the strong correlation between socioeconomic status and divorce.
      5. Frequency of attendance at religious services tends to be associated with the divorce rate: By religion, the lowest divorce rate is for Jews, then Catholics, and then Protestants.
   D. Different aspects of the life course may affect the probability of divorce.
      1. Both African-Americans and whites have a slightly increased likelihood of divorce if their families of origin were disrupted by divorce or desertion: Overall, coming from a divorced family appears to have relatively little effect on adult children's divorcing.
      2. Adolescent marriages are more likely to end in divorce than are marriages that take place when people are in their twenties.
      3. Premarital pregnancy by itself does not significantly increase the likelihood of divorce: If the pregnant woman is an adolescent, drops out of high school, and faces economic problems following marriage, the divorce rate increases dramatically.
   E. The actual day-to-day family processes may be the most important factors affecting divorce.
      1. A strong link between marital happiness and divorce appears to be true only during the earliest years of marriage: Alternatives to one's marriage and barriers to divorce appear to influence divorce decisions more than marital happiness.

Chapter 14

2. It is not clear what relation children have to divorce.
   a. One of the most significant findings in research indicates that parents of sons are less likely to divorce.
   b. Premaritally conceived children and physically or mentally limited children are associated with divorce.
3. Kitson and Sussman (1982) found that the four most common reasons given for divorce were, in descending order: personality problems, home life, authoritarianism, and differing values.

III. THE DIVORCE PROCESS 496
   A. The divorce process is not a single event, but rather a complex process: Anthropologist Paul Bohannan outlines a process made up of six "divorces:"
   B. The emotional divorce is when at least one partner begins to put emotional distance into the marriage: The heart of the marriage is missing.
   C. The legal divorce is the court-ordered termination of a marriage: Many unresolved issues of the emotional divorce may be acted out.
   D. The economic divorce entails the settlement of the property of the marriage: Alimony and child support may be required.
   E. The co-parental divorce includes the issues of child custody, visitation and support, as well as the impact of the divorce on the children: This may be the most complicated aspect of divorce.
   F. The community divorce involves changes in the social context, such as relationships with in-laws and friends.
   G. The psychic divorce is accomplished when the former spouse becomes irrelevant to one's sense of self and emotional well-being: Bohannon regards the psychic divorce as the most important element in the divorce process.

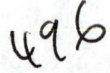

IV. MARITAL SEPARATION
   A. The crucial event in marital breakdown is not the divorce, but rather the process of marital separation: Divorce is a legal consequence that follows the emotional fact of separation.
      1. The uncoupling process usually begins as a quiet, unilateral process, as the dissatisfied person begins to turn elsewhere: The process appears to be the same for married and unmarried couples and for gay and lesbian relationships.
      2. Uncoupling does not end when the end of the relationship is announced, or even when the couple physically separate: Acknowledging that the relationship cannot be saved represents the beginning of the last stage of uncoupling.
   B. Most newly separated people do not know what to expect and many feel as if they have "lost an arm or a leg."
      1. **Separation distress**, situational anxiety caused by separation from an attachment figure, is a common experience.

2. Sooner or later the negative aspects of separation are balanced with the positive aspects, such as the possibility of finding a more compatible partner, or constructing a better life.
3. During separation distress, almost all attention is centered on the missing partner and is accompanied by apprehensiveness, anxiety, fear, and often panic.
4. Although sometimes the immediate effect of separation is not distress but euphoria, almost everyone falls back into separation anxiety.
5. As the separation continues, separation distress slowly gives way to loneliness.
6. An unexpected separation, without forewarning, is probably the most painful for the partner who is left.
7. A person goes through two distinct phases in establishing a new identity following marital separation.
   a. The transition period begins with the separation and is characterized by separation distress and loneliness: The transition period generally ends within the first year.
   b. The recovery period usually begins in the second year, when the individual has created a reasonably stable pattern of life: Emotional intensity related to the former spouse declines, yet the individual still has self-doubts.
C. Dating again presents problems such as meeting new people, being part of a singles subculture, dealing with child care, coping with strained finances, and facing sexual feelings.
   1. The functions of dating after separation/divorce include: (1) sending the message to the world that the individual is available to become someone else's partner, (2) enhancing the individual's self-esteem, and (3) initiating the individual into the singles' subculture to explore new freedoms.
   2. Several features of dating following separation and divorce differ from premarital dating: (1) Dating does not seem to be a leisurely matter; (2) Dating may be less spontaneous; (3) Finances may be strained; and (4) Separated and divorced men and women often have a changed sexual ethic.

V. CONSEQUENCES OF DIVORCE
   A. Most divorces are not contested and are settled out of court through negotiation; however, divorce is still a complex legal process involving highly charged emotions.
   B. All fifty states now have adopted **no-fault divorce**, the legal dissolution of a marriage in which guilt or fault does not have to be established.
      1. No-fault divorce has changed four basic aspects of divorce.
         a. No one is "guilty."
         b. There is no adversary process.
         c. Divorce settlements are based on equity, equality, and need.
         d. No-fault laws are intended to promote gender equality.

Chapter 14

C. Unanticipated consequences of no-fault divorce may have placed older homemakers and mothers of younger children at a disadvantage.
   1. One of the most striking differences between two-parent and single-parent families is poverty.
   2. The majority of single mothers become poor as a result of marital disruption.
   3. Husbands typically enhance their earning capacity during marriage, while wives often quit or limit their participation in the work force to fulfill family roles: This limits wives earning capacity when they reenter the work force.
   4. About a quarter of divorced women enter a spell of poverty sometime during the first five years following divorce.
D. Employment opportunities of divorced women are constrained by the necessity of caring for children.
   1. Separation and divorce dramatically change many mothers' employment patterns.
   2. Most employed single mothers are on the verge of financial disaster.
   3. Gender discrimination in unemployment and lack of societal support for child care condemn millions of single mothers and their children to poverty.
E. **Alimony** is the money payment a former spouse makes to the other to meet his or her economic needs.
   1. **Child support** is a monetary payment made by the noncustodial spouse to the custodial spouse to assist in child-rearing expenses.
   2. The Child Support Enforcement Amendments and the Family Support Act require states to deduct delinquent support from fathers' paychecks, authorize judges to use their discretion when support agreements cannot be met, and mandate periodic reviews of award levels to keep up with inflation rates.
   3. Child support awards are historically small, usually amounting to 10% of the noncustodial father's income and less than half of the child's expenses.

V. CHILDREN AND DIVORCE
A. A traditional nuclear family, merely because it is intact, does not necessarily offer an advantage to children over a single-parent family or stepfamily.
   1. Children living in happy two-parent families appear to be the best adjusted, and those from conflict-ridden two-parent families appear to be the worst adjusted: Children from single-parent families are in the middle.
   2. The key to children's adjustment following divorce is the lack of conflict between divorced parents.
   3. Telling children their parents are separating is very difficult; whether or not parents are relieved about the separation, they often feel extremely guilty.
B. Divorce involves a series of events and changes in the life circumstances of the children, not an isolated incident.
   1. Children react differently to divorce because of factors such as temperament, sex, age, and past experiences.

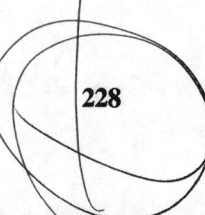

2. There appears to be a three-stage process of divorce for children:
   a. the initial stage characterized by high stress, escalated conflict, and unhappiness,
   b. the transitional stage characterized by restructuring of the family and by economic and social change, and
   c. the restabilization stage with the post-divorced family established.
C. There are six developmental tasks which children must undertake when their parents divorce:
   1. acknowledging parental separation,
   2. disengaging from parental conflicts,
   3. resolution of the loss of the familiar parental relationship, as well as their everyday routines and structures,
   4. resolution of anger and self-blame,
   5. accepting the finality of divorce, and
   6. achieving realistic expectations for later relationship success.
D. Children's responses to divorce vary by age group.
   1. Younger children may react with feelings of guilt, anger, sorrow and relief.
   2. Most children, regardless of their age, are angry because of the separation.
   3. Very young children tend to have temper tantrums while older children become aggressive.
   4. For adolescents, parental separation is traumatic.
      a. Adolescents tend to protect themselves by distancing themselves and appearing cool and detached.
      b. Adolescents are likely to be angry with both parents, blaming them for upsetting their lives.
   5. Recent research has found that marital discord may exacerbate children's behavior problems: This study reinforced a growing body of evidence showing that many problems assumed to be due to divorce are actually present before marital disruption.
   6. Regardless of the child's age, it is important that the absent parent continue to play a role in the child's life.
E. Children's adjustment can be increased by: parents discussing issues prior to the separation; continuing the child's involvement with the non-custodial parent; lack of hostility between the parents; good emotional and psychological adjustment to the divorce on the part of the custodial parents; and practicing good parenting skills.
F. The greatest damage occurs when parents use children as pawns after a divorce.

VII. CHILD CUSTODY
   A. Standards of court-awarded custody are generally based on the best interests of the child or the least detrimental of the alternatives available.
   B. In practice, the courts usually favor the mother for custody because:
      1. women usually prefer custody, men do not,
      2. tradition has given women custody, and
      3. the law is biased in that it assumes women are naturally better at caring for children.
   C. Types of Custody
      1. **Sole custody** is when the child lives with one parent, who has sole responsibility for physically raising the child and making all decisions regarding his or her **upbringing**.
         a. This is the most common custody arrangement in the U.S., accounting for 85 percent of all divorce cases.
         b. Women have traditionally been responsible for child rearing, thus, sole custody is the closest approximation to the traditional family.
         c. Many men do not feel competent, or are not perceived as competent in day-to-day child rearing responsibilities.
      2. **Joint custody**, in which both parents continue to share legal rights and responsibilities as parents, is becoming increasingly accepted, accounting for 10 percent of cases.
         a. In **joint legal custody**, the children live primarily with one parent, but both parents share in decisions regarding the children.
         b. In **joint physical custody**, children live with both parents dividing time more or less equally between the two households.
         c. Joint custody has numerous advantages, but it also requires parents to work out both the practical logistics, as well as their feelings about each other.
      3. **Split custody** splits the children between the parents, usually with girls living with their mother and boys with their father.
   D. Noncustodial Parents
      1. Noncustodial parent involvement exists on a continuum regarding caregiving, decision making, and parent-child interaction.
      2. Noncustodial parents are primarily men, many of whom suffer from the disruption or disappearance of their father role following divorce.
      3. Children tend to have little contact with the nonresidential parent: The reduced contact seems to weaken the bonds of affection.
   E. Custody Disputes and Child Stealing
      1. As many as one-third of all post-divorce legal cases involve children.
      2. About 350,000 children are abducted each year by family members in child custody disputes.

VIII. DIVORCE MEDIATION

A. **Divorce mediation** is the process in which a mediator attempts to assist divorcing couples in resolving personal, legal, and parenting issues in a cooperative manner.

B. Mediators generally have professional backgrounds in marriage counseling, family therapy or social work.

C. A primary goal of mediation is to encourage parents to view shared custody as a viable alternative.

D. Mediators try to help couples develop communication skills to negotiate with each other and suggest ways to minimize conflict.

E. Mediation is not a panacea for the difficulties of divorce — the stresses and conflicts of divorce are real and painful.

IX. READINGS AND FEATURES

A. *Other Places... Other Times: Divorce in the Chinese-American Family* proposes that the low divorce rate among Chinese-Americans reflects the lack of choices Chinese-American women have, rather than a high level of marital quality.

B. *You and Your Well-Being: Gender and Divorce Related Stressors* discusses how gender influences responses to divorce.
   1. Research indicates that divorced men experience greater emotional distress, possibly because of their more frequent social isolation.
   2. The immediate impact of divorce on women is economic.
   3. Social support is positively correlated with lower distress and positive adjustment.
   4. As with other stressors, it is often the individual's perception of the divorce itself that influences how a person adjusts to change.

C. The *Perspective: Lesbians, Gay Men and Divorce* examines the process of getting married and the consequences when lesbians and gay men divorce from a heterosexual spouse.

D. *Other Places ... Other Times: Cross-Cultural Issues in Parenting: Who Gets the Children?* explores kinship systems, patrilineal kinship and matrilineal kinship.

TEST YOUR COMPREHENSION

On the next page is a chart illustrating the effects of divorce on various members of the family. Complete the chart using readings, lectures and personal research.

Chapter 14

Chart 14

THE EFFECTS OF DIVORCE ON FAMILY MEMBERS	
Member(s)	**Effects**
The whole family	
The husband/father	
The wife/mother	
Pre-school children	
School children	
Adolescents	
The grandparents	

SELF QUIZZES

How well do you know this material? Test your understanding of the reading assignments by answering the following sample questions.

PART I - Multiple Choice: Chose the most correct response.

____C____ 1. Various factors which researchers have found to be related to divorce include all but which of the following?
   a. the shift from an agricultural society to an industrial society
   b. individualism taking priority over family
   c. a devaluation of marriage by our society
   d. frequency of attendance at religious services
   e. ethnicity

____e____ 2. Which of Bohannan's "divorces" does he consider to be the most important?
   a. the emotional divorce
   b. the legal divorce
   c. the economic divorce
   d. the co-parental divorce
   e. the psychic divorce

____e____ 3. Which of the following is not true regarding divorce?
   a. Except for the death of one's spouse, divorce is the largest stress-producing event in a person's life.
   b. Social scientists are increasingly viewing divorce as one path in the normal family life cycle.
   c. The crucial event in a marital breakdown is not the divorce, but the emotional fact of separation.
   d. Divorce is an important element of the contemporary American marriage system because it reinforces the significance of emotional fulfillment in marriage.
   e. All of the above are true.

____b____ 4. All but which one of the following are true regarding "uncoupling"?
   a. The initiators often ponder questions fundamental to their identity.
   b. It is an intentional turning away from the spouse.
   c. It creates division in the relationship.
   d. Both the initiator and the partner try to cover up the seriousness of the dissatisfaction.
   e. Uncoupling does not end when the couple physically separates.

Chapter 14

___b___ 5. Which of the following best illustrates separation distress?
   a. It is a condition of depression due to the loss of financial security that the spouse provided.
   b. It is a condition of stress, anxiety and anger when we lose an "attachment figure" with whom we associate comfort and security.
   c. It is a condition of unhappiness that children experience when they are told their parents are going to divorce.
   d. It is a condition of anger due to the resentment each still retains for the other.
   e. all of the above

___b___ 6. Which of the following is not true about dealing with separation?
   a. Almost everyone suffers separation distress when a marriage breaks up.
   b. It takes between six and twelve months to recover fully from a marital breakup.
   c. Most people are surprised by how long the recovery takes.
   d. Friends often splinter, and may choose one or the other of the former partners for continuing friendships.
   e. Men and women react differently to separation.

___d___ 7. Which of the following is not true of divorce in adults?
   a. Whether a person had warning and time to prepare for a separation affects separation distress.
   b. Divorced couples experience greater economic stress.
   c. Divorced men experience greater emotional distress and report more suicidal thoughts than do women.
   d. Sexual life is less important for divorced men and women.
   e. Divorce serves as a recycling mechanism, giving people a chance to improve their marital situation by marrying again.

___b___ 8. Which of the following is not characteristic of dating after divorce?
   a. It may be less spontaneous than premarital dating.
   b. People dating after divorce are more hesitant to engage in sexual activity.
   c. It is not as leisurely a matter as in premarital dating.
   d. It signals availability to become someone else's partner.
   e. It is an opportunity to enhance self-esteem.

___d___ 9. Which of the following characterizes no-fault divorce?
   a. It charges the spouse with adultery or desertion.
   b. It charges the spouse with nonsupport and cruelty.
   c. It assumes that one partner is clearly responsible for the marital breakup.
   d. Neither party is found guilty, the marriage is declared unworkable and is dissolved.
   e. It assumes an adversary process and also assumes that women need to be supported by men.

Coming Apart: Separation and Divorce

___e___ 10. A problem of no-fault divorce for women is that
  a. the law assumes that divorced women can become immediately self-sufficient economically.
  b. many women who have spent years as homemakers have outdated experience, few skills, and no seniority.
  c. women may have enhanced their husband's earning capacity at the expense of their own.
  d. they and their children are systematically impoverished by the no-fault divorce laws.
  e. all of the above

___c___ 11. Which of the following is not a developmental task of children of divorce?
  a. disengaging from parental conflicts
  b. acknowledging parental separation
  c. being an emotional support for the custodial parent
  d. resolution of self-blame
  e. All of the above are developmental tasks of children of divorce.

___e___ 12. Which of the following behaviors is(are) characteristic of children when dealing with a divorce?
  a. The children may blame the custodial parent for making the other parent leave.
  b. Younger children may be fearful that the custodial parent will abandon them.
  c. Preadolescent children may regress to immature behavior, bed wetting or becoming excessively possessive.
  d. Some children may have temper tantrums and may be more aggressive in their play.
  e. all of the above

___a___ 13. Which of the following is not true regarding the custody of children?
  a. Alimony and child support usually cover child care expenses.
  b. Women usually get custody of children, partly because more women seek custody.
  c. The law reflects a bias that assumes women are naturally better able to care for children.
  d. Recently, increasing numbers of fathers have been gaining custody of their children following divorce.
  e. Custody may be sole, joint or split.

___e___ 14. Goals of divorce mediation include all but which one of the following?
  a. encouraging divorcing parents to see shared parenting as a viable alternative
  b. helping couples develop communication skills
  c. helping parents determine whether their demands are based on anger
  d. helping parents clarify personal goals
  e. All of the above are goals of mediation.

235

Chapter 14

PART II - True/False

__T__ 1. Before 1974, the majority of marriages ended through death rather than divorce.

__F__ 2. Researchers today view divorce as deviant.

__F__ 3. The degree of interaction between individuals and the larger community has relatively little impact on their divorce rate.

__T__ 4. Coming from a divorced family appears to have relatively little effect on adult children's divorcing.

__F__ 5. Regardless of years married, there is a strong link between marital happiness and divorce.

__T F__ 6. Extramarital affairs have been found to be the most common reason for divorce.

__T__ 7. The co-parental divorce may be the most complicated aspect of divorce.

__T F__ 8. The crucial event in the marital breakdown is the actual divorce more than the separation.

__T F__ 9. Uncoupling is a malicious turning away from one's partner.

__T F__ 10. It is possible to strongly miss someone after a divorce, even if the couple had been unhappy for years. True

__F T__ 11. It seems to take about two years for children to become emotionally and socially integrated into a post-divorce family.

__T F__ 12. A first date after years of marriage and subsequent months of singlehood evokes some of the same emotions felt by inexperienced adolescents.

__T__ 13. Most employed single mothers are on the verge of financial disaster.

__F__ 14. Studies indicate that no matter how much conflict existed in the home, it is better for children to have both parents than to have the family split up by divorce.

__T__ 15. The age of the children has a significant effect upon how they will handle the divorce of their parents.

__F__ 16. Adolescents are very perceptive and almost none appear to be surprised when their parents announce that they plan to divorce.

__T__ 17. Using children as pawns to maintain an attachment between oneself and an ex-spouse is one of the greatest dangers to children's post divorce adjustment.

__T F__ 18. A study of fathers a year after divorce found greater satisfaction with custody arrangements among those who litigated rather than mediated.

236

DISCUSS BRIEFLY

What are some of the demographic factors that affect our divorce rates?

SELF-DISCOVERY

Sometimes one can hear comments such as "when people get married today, they don't need to make a commitment because it is so easy to get a divorce." Comment on how you feel about both parts of this statement.

"A good divorce is better than a bad marriage." What do you think?

Should people stay married "just because of the kids?"

What are some reasons you personally might consider a divorce?

Chapter 14

MINI-ASSIGNMENT I

If it is possible in your area, attend a divorce hearing. Describe the process. What was it like? What insights did viewing the process give you?

MINI-ASSIGNMENT II

Talk to several people who have recently experienced a divorce. Did their experiences seem to support the text's findings? If so, in what ways? What additional insights did you gain from these discussions?

KEY TO SELF QUIZZES

Multiple Choice

1. c
2. e
3. e
4. b
5. b
6. b
7. d
8. b
9. d
10. e
11. c
12. e
13. a
14. e

True/False

1. T
2. F
3. F
4. T
5. F
6. F
7. T
8. F
9. F
10. T
11. F
12. T
13. T
14. F
15. T
16. F
17. T
18. F

SUGGESTED READINGS

For related readings, see pages 520 - 521 in the text.

CHAPTER 15
New Beginnings: Single-Parent Families and Stepfamilies

MAIN FOCUS

Chapter Fifteen examines single-parent families, binuclear families, remarriage, and stepfamilies. It discusses the patterns of individuals following divorce as they adjust, establish new relationships, remarry, and begin to cope with stepfamilies and step-relationships.

GOALS OF THIS CHAPTER

To demonstrate mastery of this chapter, you should be able to:

1. Know and describe the problems and characteristics of single-parent families.

2. Explain the binuclear family in terms of its complexity and its subsystems.

3. Discuss remarriage in terms of courtship, its characteristics, marital satisfaction, and stability.

4. Understand who is most likely to remarry, how second marriages differ from first ones, and the success rates of remarriages.

5. Describe how the stepfamily is structurally different than first-marriage families, and the problems facing men, women and children in their step roles.

6. Explain the developmental stages of stepfamilies.

7. Understand the sources of conflict in stepfamilies as well as appreciate stepfamily strengths.

8. Describe the phenomenon of single-parent families among African-American adolescents.

9. Discuss ways to create happy stepfamilies.

Chapter 15

KEY TERMS AND IDEAS

The following terms, ideas, and concepts are listed in the order in which they appear in Chapter Fifteen and in the outline. Be sure that you understand and can define each of the following:

single-parent families
stepfamilies
blended families
remarriage
binuclear families

CHAPTER FIFTEEN OUTLINE

I. INTRODUCTION TO THE CHAPTER
   A. The 1990's mark the definitive shift from a traditional marriage and family system to a pluralistic family system.
   B. The pluralistic family system consists of three major types of families: (1) intact nuclear families; (2) single-parent families; and (3) stepfamilies.
      1. Chances are more than two out of three that an individual will divorce, remarry, or live in a single-parent family or stepfamily as a child or parent sometime during his or her life.
      2. Remarriage is as common as first marriage.
      3. Nearly one-fourth of all families are currently single-parent families.
      4. Over 2.3 million households have stepchildren living with them, and over a third of all children can expect to live in a stepfamily at some time during their childhood.
      5. Instead of seeing single-parent and stepfamilies as "deviant," researchers are reevaluating them as normal; the focus becomes one of examining if the family is fulfilling its functions.

II. SINGLE PARENT FAMILIES
   A. **Single-parent families** are families consisting of one parent and one or more children: The parent can be either divorced or never married.
   B. Single-parent families are the fastest growing family form in the United States: Between 1970 and 1995 the percentage of single-parent families nearly doubled.
   C. Single-parent families today tend to be created by marital separation, divorce, or births to unmarried women rather than by widowhood.
      1. Divorced or unmarried single mothers receive less social support compared to widowed single mothers.
      2. Eighty-seven percent of single-parent families are headed by women.

3. Ethnicity is a significant demographic factor (In 1995, 64% of African-American children lived in single-parent families, compared to 25% of all white children and 36% of Hispanic children).
4. Most single-parent families must deal with poverty and are under constant economic stress trying to make ends meet.
5. Single-parent families rely on a variety of household arrangements and show great flexibility in dealing with child care and housing problems.
6. Single-parent families are often a transitional state between two marriages.
7. For many, especially single women in their thirties and forties, single-parenting has become a more intentional and less transitional life style.
8. There may be 2.5 to 3.5 million lesbian and gay single parents.

D. Of children born in 1980 who will live in a single-parent family, white children can expect to spend about 31% of their childhood in single-parent families, African-American children about 59%.
1. Children provide stability and company for the single parent, but do not counteract the loneliness he or she may feel.
2. The change in family structure also changes the way the parenting adult relates to the children; with the mother becoming closer to her children, more egalitarian family situations, more responsibility, and children gaining power in determining the rules.

E. Characteristics of successful single parents include:
1. acceptance of responsibilities and challenges of single parenthood,
2. parenting as first priority,
3. consistent, nonpunitive discipline,
4. emphasis on open communication,
5. fostering individuality that is supported by the family,
6. recognition of the need for self-nurturance, and
7. dedication to rituals and traditions.

F. Family strengths identified by Richards and Schmeige as associated with successful single parenting include:
1. parenting skills including the ability to assume some of the roles and attributes of the absent parent,
2. personal growth — development of a positive attitude, and feeling success and pride in overcoming obstacles,
3. good communication which develops trust and a sense of honesty,
4. management of daily family activities, and
5. developing the ability to become financially self-supporting and independent.

III. BINUCLEAR FAMILIES
A. The binuclear family is a post-divorce family system with children, the original nuclear family divided in two.
1. The binuclear family consists of two nuclear families, the maternal nuclear family headed by the mother (the ex-wife), and the paternal nuclear family headed by the father (the ex-husband).

Chapter 15

    2. Both single-parent and blended families are forms of binuclear families.
- B. The binuclear family may be extremely complex and ambiguous.
- C. Researchers divide the binuclear family into the following five subsystems.
    1. The former spouses subsystem — continuing because of parental responsibilities
        a. Issues that former spouses deal with include anger and hostility, conflict related to children and parenting, shifting roles and relationships, and incorporating step-relatives after remarriage.
        b. Former spouses who are able to separate parenting from personal issues may form effective co-parenting relationships.
    2. The remarried couple subsystem — they may have physical custody of children from their first marriages, and may have problems coping with former spouses.
    3. The parent-child subsystem — requiring adjustments for biological parents and stepparents.
    4. The sibling subsystem — in which stepsiblings and half-siblings may have to contend with each other for parental affection, toys, attention, physical space, and dominance.
    5. The mother/stepmother-father/stepfather subsystem — the relationship between new spouses and former spouses.

IV. REMARRIAGE
- A. **Remarriage** is a marriage in which one or both partners have been previously married: Nearly half of all marriages in the United States involve at least one partner who has been previously married.
- B. Characteristics affecting remarriage include:
    1. a man's or woman's age at the time of separation (the greatest individual factor affecting remarriage),
    2. ethnicity, and
    3. the presence of children. (Although children from earlier marriages traditionally have been thought to decrease the likelihood of remarriage, the evidence is mixed.)
- C. Courtship norms after separation or divorce are not clear in our society, and thus courtship may be plagued with uncertainty about what to expect or how to act.
    1. If neither person has children, the courtship process is similar to those who have never married; memories of the earlier marriage exist as a model, both negative and positive.
    2. Single parents often lack leisure time and money to participate in the singles' world of dating.
        a. The decision to go out at night may lead to guilt feelings about the children.
        b. The single parent must look at a potential partner as a potential parent.
        c. The children may resent the person their parent is dating as an "intruder" and try to sabotage the new relationship.
        d. The issue of permitting someone to spend the night becomes an important symbolic act.

D. Remarriage is considerably different from first marriages because the partners are experiencing commitment in the midst of change, they are at different stages in the life cycle, they have different expectations, and remarriage often involves the creation of stepfamilies.
E. People appear to have as much marital satisfaction in remarriages as other people who are in first marriages.
   1. Remarried couples are more likely to divorce, especially if stepchildren are present.
      a. Persons who remarry after divorce are more likely to use divorce as a way of resolving an unhappy marriage.
      b. Remarriage is an "incomplete institution" and does not receive the same family and kin support as first marriage.
      c. Remarriages are subject to stresses nor present in first marriages.
   2. According to Gagnon and Coleman, the presence of stepchildren is a major contributor to the higher divorce rate in stepfamilies, compared to childless remarried couples.

V. STEPFAMILIES
   A. **Stepfamilies** are families in which one or both partners have children from a previous marriage or relationship: Stepfamilies are sometimes called **blended families**.
   B. Stepfamilies work differently and provide different satisfactions and challenges than intact families.
   C. Six structural characteristics of stepfamilies make them different from the traditional first-marriage families.
      1. Almost all members have lost an important primary relationship.
      2. One biological parent lives outside the current family.
      3. The relationship between a parent and his/her children predates the relationship between the new partners.
      4. Stepparent roles are ill-defined.
      5. Many children in stepfamilies are also members of the non-custodial parent's household, which may have a second set of rules.
      6. Children in stepfamilies have at least one extra pair of grandparents.
   D. The seven developmental stages of stepfamilies found in the text represent a process which each individual experiences differently.
      1. Early stages include: fantasy, immersion, and awareness.
         a. In the fantasy stage, parents hope for a wonderful relationship while children tend to have quite opposite fantasies.
         b. During the immersion stage, reality replaces fantasy.
         c. The awareness stage is when family members attempt to become aware of and identify the feelings they are experiencing.
      2. The middle stages of mobilization and action involve changes in the emotional structure of the family as a whole.
         a. In the mobilization stage, family members recognize differences, conflict becomes more open, stepparents begin to take a stand, and the family begins to integrate the stepparent into its functioning.

b. In the action stage, the family begins to take major steps to reorganize itself into a stepfamily: It creates new norms and family rituals.
  3. The later stages of contact and resolution involve solidifying the stepfamily.
     a. In the contact stage, family members make intimate contact with each other: The **stepparent role** emerges as an individually created role.
     b. In the resolution stage, the stepfamily is solid and no longer requires the close attention and work of the middle stages.
  4. It takes most stepfamilies about seven years to complete the developmental process.
  5. Becoming a stepparent is a slow process that moves in small ways to transform strangers into family members.
E. Most people go into stepfamilies expecting to recreate the traditional nuclear family: The hardest adjustment is realizing that the two are different and that being different does not make stepfamilies inferior.
  1. Women in stepfamilies want to make up to the children for the divorce, create a happy family, prove they are not wicked stepmothers, love their stepchild instantly, and be loved instantly in return.
  2. Men in stepfamilies may try to be superdads to their own children, feel guilt and confusion in their new families, and experience conflict related to discipline and fitting into the stepfamily.
F. Conflict in stepfamilies exists, as it does in all families, but with special problems:
  1. Favoritism may exist along kinship lines.
  2. Divided loyalty between parents places great stress upon children.
  3. Discipline is especially difficult to handle if the child is not one's biological offspring.
  4. The problem of allocating money, goods, and services is often difficult in stepfamilies.
     a. In the one-pot pattern of resource distribution, families pool their resources and distribute them according to need.
     b. In the two-pot pattern of resource distribution, resources are distributed by biological relationship, with need being secondary.
G. The strengths of stepfamilies need to be emphasized.
  1. Stepfamilies are able to fulfill traditional family functions.
  2. A binuclear single-parent, custodial family, or noncustodial family may provide more companionship, love, and security than the particular traditional nuclear family it replaces.
  3. Stepfamilies can potentially offer children a number of benefits which can compensate for the negative consequences of divorce.
     a. Children gain multiple role models.
     b. Children gain greater flexibility.
     c. Stepparents may act as a source of support and information.
     d. Children may gain additional siblings.

       e. Children gain an additional extended kin network which may become important and loving.
       f. A child's economic situation is often improved.
       g. Children may gain parents who are happily married.

VI. READINGS AND FEATURES
   A. *Single-Parent Families Among African-American Adolescents* explores the factors around African-American teens being twice as likely to be sexually active as whites and their higher birth rates.
      1. The stereotypical image of the adolescent African-American mother reflects a minority of cases.
      2. The most pressing needs of teen mothers are health care and education.
      3. Rather than the assumption that teenage fathers are irresponsible, adolescent fathers more typically remain physically or psychologically involved throughout the pregnancy.
   B. In the feature *Parental Images: Biological Parents versus Stepparents*, stereotypes and feelings related to how children perceive their parents and stepparents are examined.
   C. *You and Your Well-Being: Hints for Creating Happy Stepfamilies* proposes that the following steps can help stepfamilies:
      1. Be honest with yourself and your partner before marriage.
      2. Acknowledge that changes in life style will occur for everyone involved.
      3. Discuss similarities and expect differences in childrearing.
      4. Invite and encourage children to share their feelings.
      5. Be patient.
      6. Be persistent.
      7. Acknowledge the need for support.
      8. Make the marriage the primary relationship.
      9. Seek help, if needed.

Chapter 15

TEST YOUR COMPREHENSION

Below is a diagram illustrating the roles and relationships of two blended families. In the boxes, discuss each relationship marked with an arrow, using readings, lectures and personal experience or research. A few areas may not have been discussed within the text.

Diagram 15.

FAMILY DYNAMICS OF TWO STEPFAMILIES

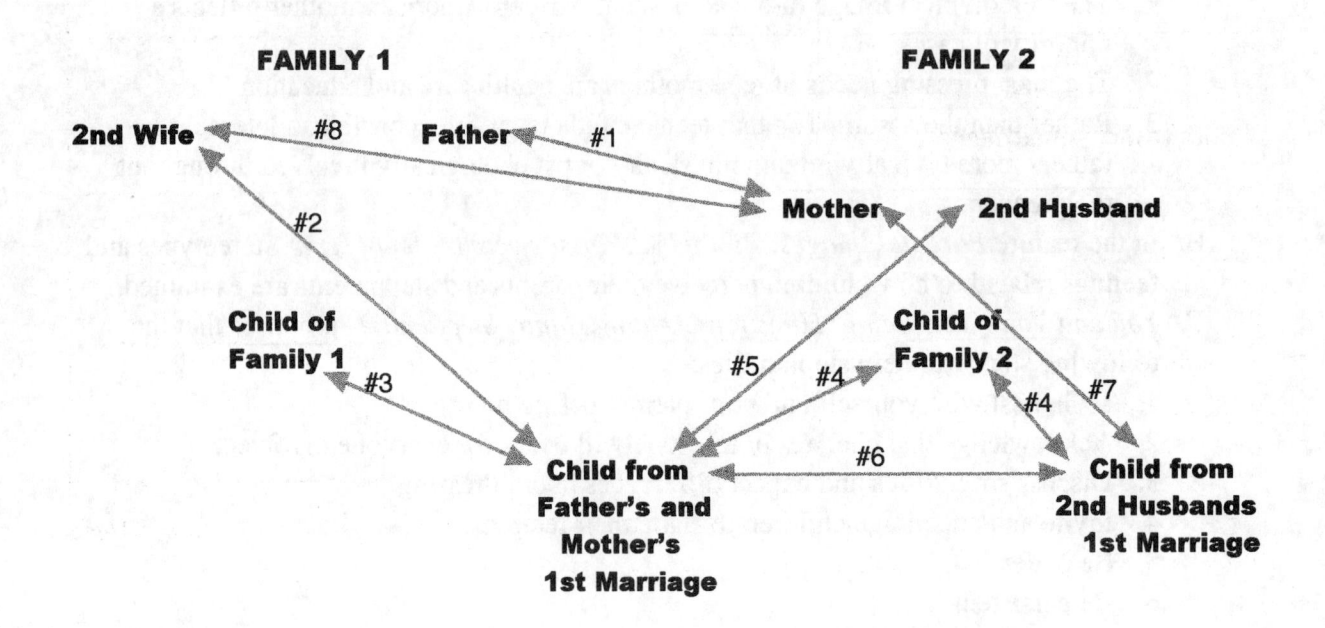

FAMILY 1 — FAMILY 2

2nd Wife ← #8 — Father ← #1 → Mother ↔ 2nd Husband

#2, Child of Family 1, #3

Child of Family 2, #5, #4, #7

Child from Father's and Mother's 1st Marriage ↔ #6 ↔ Child from 2nd Husband's 1st Marriage

Relationship	Description of Relationships
Relationship #1 between ex-spouses	
Relationship #2 between 2nd wife and husband's child from 1st marriage	
Relationship #3 between child of Family 1 and father's child from his 1st marriage	

New Beginnings: Single-Parent Families and Stepfamilies

Relationship	Description of Relationships
Relationship #4 between child of 2nd marriage and half-siblings from parent's 1st marriages	
Relationship #5 between 2nd husband and his stepchild from his wife's 1st marriage	
Relationship #6 between mother's child and 2nd husbands's child both from 1st marriages	
Relationship #7 between stepmother and stepchild	
Relationship #8 between 1st wife and 2nd wife	

This chart does not illustrate other existing family relationships. Mark and number them, using red pen or pencil, and list those possible relationships below:

How many possible relationships did you find?

Imagine the addition of other children, grandparents, aunts, and uncles, and you can see how complicated stepfamilies can be! (If you would like, add grandparents with lines detailing relationships in another color.)

Chapter 15

SELF QUIZZES

How well do you know this material? Test your understanding of the reading assignments by answering the following sample questions.

PART I - Multiple Choice: Choose the most correct response.

__a__ 1. Which of the following is not true of single parents?
   a. Welfare payments place the majority of single-parent families above the poverty line.
   b. White single mothers are more likely to be divorced than their African-American or Latino counterparts.
   c. There may be as many as 2.5 to 3.5 million lesbian and gay single parents.
   d. Single parenting is usually a transition period.
   e. Ethnicity is an important demographic factor in single-parent families.

__c__ 2. Single parents are likely to change their discipline standards by
   a. being increasingly strict.
   b. being increasingly authoritarian.
   c. being more willing to compromise.
   d. being more distant.
   e. none of the above

__c__ 3. Olsen and Haynes found all but which of the following to be characteristic of successful single parents?
   a. parenting as first priority
   b. emphasis on open communication
   c. unity stressed over individuality
   d. dedication to rituals
   e. consistent, nonpunitive discipline

__b__ 4. Which of the following is not true regarding the binuclear family?
   a. The binuclear family is a post-divorce family system with children.
   b. Divorce ends a marriage and a family.
   c. Ex-husbands and ex-wives may continue to relate to each other and to their children.
   d. In single-parent families headed by women, the paternal family component may be minimal.
   e. It may be the most complex family system in America.

248

New Beginnings: Single-Parent Families and Stepfamilies

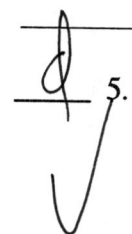

5. Which of the following is not listed in your text as a subsystem in the binuclear family?
   a. the former spouse subsystem
   b. the remarried couple subsystem
   c. the parent/child subsystem
   d. the grandparent/parent/child subsystem
   e. the mother/stepmother-father/stepfather subsystem

6. For single parents, dating presents problems such as
   a. feeling guilty in dating and leaving the children at home.
   b. looking at a potential partner as a potential parent.
   c. whether to permit someone to spend the night when the children are present.
   d. revealing to the children that their parent is sexual and may have relationships apart from them.
   e. all of the above

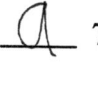

7. Which of the following is not likely to be true of remarriages?
   a. Remarriages are more likely to be based on romance and sexual desire than are first marriages.
   b. Remarriages usually involve great hope, but also fear of the past.
   c. Divorced people are at a different stage in their life cycle than when they first married.
   d. Divorced people have different expectations of what they want out of their new marriage.
   e. Remarriage is an "incomplete institution."

8. Which of the following is likely to be true of a second marriage?
   a. Remarrieds are likely to report their marriages less happy than those who are in first marriages.
   b. Partners in remarriage are more willing to work hard to keep their marriage together so as to avoid the pain of divorce again.
   c. Remarried couples are more likely to divorce than couples in first marriages.
   d. Remarriages receive strong support and clear norms from society.
   e. Remarriages that involve children usually have few adjustment problems.

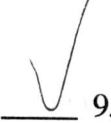

9. Stepfamilies are structurally different from first-marriage nuclear families because
   a. almost all members have lost an important primary relationship.
   b. one biological parent lives outside the current family.
   c. the relationship between the parent and the child pre-dates the relationship with the stepparent.
   d. the roles of stepparents are poorly defined.
   e. all of the above

Chapter 15

___10. The early developmental stages of becoming a stepfamily includes all but which of the following?
   a. contact
   b. immersion
   c. fantasy
   d. awareness
   e. All of the above occur in the early stages.

___11. Women in stepfamilies generally expect to
   a. make up to the children for the divorce.
   b. create a happy, close-knit family.
   c. prove they are not wicked stepmothers.
   d. receive love from the stepchildren instantly.
   e. all of the above

___12. Which of the following is not characteristic of stepparenting?
   a. Stepmothers tend to experience greater stress in stepfamilies than stepfathers.
   b. A critical factor in a man's stepparenting is whether he has children of his own.
   c. The stepmother frequently assumes the role of disciplinarian.
   d. Stepfathers tend not to be as involved as stepmothers.
   e. All of the above are characteristic of stepparenting.

___13. Which of the following is the least likely to be a conflict in stepfamilies?
   a. personal conflicts
   b. issues surrounding favoritism
   c. arguments about discipline
   d. divided loyalties
   e. problems allocating money

___14. Which of the following can be viewed as a benefit of stepfamilies?
   a. They are quite capable of fulfilling traditional family functions.
   b. They may be considerably better than the family they replace.
   c. Children may gain additional family members and grandparents.
   d. Children may gain parents who are happily married.
   e. all of the above

___15. Factors contributing to African-Americans having a higher teenage pregnancy rate than whites include all but which of the following?
   a. African-Americans are less likely to abort an unintended pregnancy.
   b. The gender ratio of available African-American men to women put African-American women at a disadvantage to marry.
   c. African-Americans tend to use contraception less consistently than whites.
   d. Three generation households are more common among whites than African-Americans.
   e. All of the above are true.

New Beginnings: Single-Parent Families and Stepfamilies

PART II - True/False

_T_ 1. Remarriage is as common as first marriage.

_T_ 2. Almost one out of every four families is a single-parent family.

_T_ 3. Approximately half of all children born in the 1990's will spend part of their childhoods in single-parent families.

_T_ 4. Many of the problems of single-parent families and stepfamilies lie in the stigma attached to them and their lack of support by the larger society.

_F_ 5. Seventy percent of single-parent families are headed by women. ✗ 87%

_F_ 6. A child's need for the parent compensates for any loneliness a single parent might experience.

_F_ 7. Many women find that being a single parent is even more hopeless than an unhappy marriage.

_F_ 8. Even when there are children, divorce usually ends all relationships between the two former spouses.

_F_ 9. In remarried couple systems, typical marital issues such as power and intimacy tend to be less intense than in first marriages.

_T_ 10. Remarriage is probably more difficult for children than for parents.

_T_ 11. Nearly half of all marriages in the United States are marriages in which at least one partner has been previously married.

_F_ 12. After a person has been divorced, he/she is unlikely to remarry for fear of another failure.

_F_ 13. Remarriage courtships are short. Long

_T_ 14. Remarriage has often been regarded as the pathway to well-being.

_F_ 15. Remarried couples are less likely to divorce than first married couples.

_T_ 16. During the mobilization stage of stepfamilies, members "map" the territory of the family.

_T_ 17. Both the stepmother and the stepfather are likely to have extremely unrealistic expectations about their roles in the newly established family. True

251

Chapter 15

___A___ 18. Healthy stepfamilies are less cohesive and more adaptable than healthy intact families.

___A___ 19. Stepfamilies may provide multiple role models and more flexibility for children.

___A___ 20. The percentage of single-parent families increases significantly for African Americans.

DISCUSS BRIEFLY

1. How do remarriages differ from first marriages? Are they likely to be stronger?

2. What factors seem to contribute to the success in stepfamilies? What special sources and forms of conflict are likely in a stepfamily? What are the strengths of stepfamilies?

3. What are some of the factors of being single and being a parent which make it difficult to establish new relationships?

4. Outline the seven stages in the developmental process in stepfamily formation.

5. What are some ways stepfamily life can be beneficial to children?

Chapter 15

KEY TO SELF QUIZZES

Multiple Choice

1. a   11. e
2. c   12. e
3. c   13. a
4. b   14. e
5. d   15. d
6. e
7. a
8. c
9. e
10. a

True/False

1. T   11. T
2. T   12. F
3. T   13. T
4. T   14. T
5. F   15. F
6. F   16. F
7. F   17. T
8. F   18. T
9. F   19. T
10. T 20. T

SUGGESTED READINGS

For related readings, see page 557 in the text.

CHAPTER 16

Marriage and Family Strengths

MAIN FOCUS

Chapter Sixteen examines marital strengths in comparison to family strengths, the family as a process, characteristics of strong families, and cohesiveness versus adaptability. It also looks at family form and ethnic identity, kin, friendships, and the family in the community.

GOALS OF THIS CHAPTER

To demonstrate mastery of this chapter, you should be able to:

1. Recognize marital strengths and family strengths and discuss the relationship between them.

2. Explain the importance of the work of families.

3. Recognize and explain the characteristics of strong families.

4. Describe the role of commitment and communication as well as the importance of affirmation, respect, and trust in establishing and maintaining strong families.

5. Explain the significance of role-modeling, responsibility, self-respect, moral values and spiritual wellness in strong families.

6. Describe the importance of tradition, rituals, and family history (especially for ethnic families) in establishing continuity with the past and in hopes for the future.

7. Explain how the capacity to deal effectively with family crisis and the ability to seek help operate in strong families.

Chapter 16

8. Acknowledge the importance of play and leisure time in strong families.

9. Explain cohesiveness and adaptability as central dimensions of family functioning.

10. Explain family strengths in various family forms and ethnic identities.

11. Understand the importance of kin, community, and personal friendships in supporting individuals and families.

12. Describe the various traditions and rituals in modern American life.

13. Explain cohesion and adaptability in stepfamilies in comparison to intact families without divorce.

KEY TERMS AND IDEAS

The following terms, ideas, and concepts are listed in the order in which they appear in Chapter Sixteen and in the outline. Be sure that you understand and can define each of the following:

family strengths	bondedness	coalitions
family wellness	boundaries	adaptability
cohesion		

CHAPTER SIXTEEN OUTLINE

I. INTRODUCTION TO THE CHAPTER
   A. The vital place of family has never been seriously questioned: Working with the family is one of the first solutions suggested for virtually all social problems.
   B. How well a family is able to accomplish basic tasks of the family depends on a number of characteristics and abilities.

II. MARITAL STRENGTHS
   A. Marriage may be seen as a forum for negotiating the balancing between the desire for intimacy and the need to maintain a separate identity through interpersonal competence.
   B. Marital strengths versus family strengths must be examined, as many of the traits of healthy marriages are also found in healthy families.
      1. Childfree couples generally have more time for each other and substantially less psychological, economic, and physical stress.
      2. Many of our marital skills probably develop alongside our family skills.
      3. The relationships of families with children generally have greater stability because the emotional cost of a breakup is much greater when children are present; in addition, children help fulfill our need for intimacy.

C. David Mace cites the essential aspects of successful marriage as commitment, communication, and the creative use of conflict.
   1. Numerous studies show a strong correlation between a couple's communication patterns and marital satisfaction.
   2. Commitment is defined in terms of ongoing growth, involving the willingness and ability to work: Commitment to the sexual relationship within marriage appears to be an important aspect of marital strength.
   3. Commitment involves give-and-take in order to be together and nurture the marriage relationship.
   4. Commitment to success is an essential component in the formation of strong marriages and strong families.

III. FAMILY STRENGTHS
   A. **Family strengths** are those characteristics that contribute to a family's satisfaction and its perceived success as a family.
   B. An important part of the work of being a family and building family strengths is the identification of family goals: Family goals are unique to each family.
   C. Perfection in families exists as an ideal: Family quality can be seen as a continuum.
   D. Family quality varies over the family life cycle: The family as a process illustrates the fact that all families are changing and growing constantly.
      1. The overall cohesiveness of the family is severely tested at times; although it often emerges stronger, it may have experienced periods of distrust, disorder and unhappiness.
      2. Families are idiosyncratic; each is different from all others.
   E. Family is the irreplaceable means by which most of the social skills, personality characteristics, and values of individual members of society are formed.
      1. The climate of the family determines the way and the milieu in which care giving is offered: This affects the social, psychological, and spiritual characteristics of the individual family members.
      2. Family life professionals: (1) attempt to become aware of family problems and to devote research and service where needed; (2) strive to take the time and the opportunity to recognize and celebrate the tremendous contribution families make to our society; and (3) try to educate the public that strong families do not just happen, but rather involve work.
   F. Research shows that strong families share family strengths that create a sense of positive family identity.
   G. Commitment involves the promotion of growth of other family members: It involves working on behalf of one's family — the best for each person.
      1. Commitment is a prevailing characteristic in strong families of all forms.
      2. Commitment to the family involves the participation of family members in a world view that encompasses more than self-centered interest.
      3. True commitment is revealed and renewed in our actions.

H. Affirmation, respect, and trust are essential to family health.
   1. Supporting others in our family and being supported and affirmed in return are important in maintaining a feeling of satisfaction and well-being.
   2. According respect to our family members for their uniqueness and differences and encouraging members to develop their individuality is also important.
   3. Criticism, ridicule, and rejection undermine self-esteem.
   4. The establishment of trust that family members can be relied on encourages the development of self-confidence and a sense of responsibility for oneself and others.
   5. Parental role-modeling is a crucial factor in the development of qualities that ensure personal psychological health and growth.
I. Our tone of voice, body language, eye contact, silences, a touch, or a gift are all forms of communication.
   1. In strong families, communication is direct.
   2. Strong families talk a lot, trust one another, and are good listeners.
   3. In times of conflict, strong families keep communication focused on the issues rather than the personalities of those involved.
   4. Communication has been described as a huge umbrella that covers all that transpires among human beings.
   5. Communication facilitates other family strengths.
J. Responsibility, morality, and spiritual orientation are important for healthy families.
   1. The acquisition of responsibility is rooted in a sense of self-respect and an appreciation of the interdependence of people.
      a. Successful families realize the importance of delegating responsibility and of developing responsible behavior.
      b. Parental acknowledgment of a job well done goes a long way toward building responsibility in children.
      c. Healthy families know the importance of allowing children to make their own mistakes and face the consequences.
   2. Healthy families develop a sense of right and wrong, a moral code, in their children based on a firm conviction that the world and people around us must be valued and respected.
   3. Families with a spiritual orientation see a larger purpose for their family than simply their own maintenance and self-satisfaction.
   4. Spirituality gives meaning, purpose, and hope.
K. Families are enhanced through traditions and having a sense of family history.
   1. By traditions and rituals, families find a link to the past and a hope for the future.
   2. Strong families often have a sense of family history, strengthening a sense of connection to its roots.

L. Research consistently identifies the capacity to deal effectively with family crises as a characteristic of strong families: Members of a strong family unite to face the challenges of a crisis.
   1. The cumulative effect of other family strengths enables strong families to deal with crises.
   2. Strong families are able to accept changes resulting from crises and to see possibilities for growth in them.
M. Effective crisis management is associated with the family's ability to be open to resources outside itself.
   1. Strong families acknowledge their vulnerabilities.
   2. Strong families' experiences of interdependence within the family better equip them to recognize the interdependence among families and community.
N. Making time for family is an interlocking trait that brings together other characteristics found in strong families: It expresses commitment to the family.
   1. Spending time together is necessary to develop adequate communication and to build cohesion.
   2. Healthy families give play and leisure time high priority.
O. Having a **family wellness** orientation means making a conscious decision to live our lives in ways that move us toward optimal health in physical, emotional, intellectual, spiritual, and social dimensions.
   1. Wellness is positive, proactive, and focuses on being healthy and whole.
   2. Families oriented toward wellness take advantage of educational opportunities that help them gain perspectives on family developmental processes.
P. **Cohesion** and adaptability are central dimensions to family functioning.
   1. Family cohesion, or closeness, refers to the emotional bonding that family members have toward one another.
   2. Elements which contribute to a family's cohesion include emotional bonding (**bondedness**); **boundaries**; **coalitions**; sharing time, space, interests, recreation, friends; and decision making.
   3. An understanding of how the individual development of children and parents influence the nature of cohesion at different stages in the family life cycle is important in strong families.
   4. Family **adaptability** is the ability of the marital/family system to change in response to situational and developmental stress.
      a. Six dimensions appear to be critical to family adaptability: leadership, assertiveness, discipline, negotiation, roles, and rules.
      b. Balance offers the most potential for the adaptability that leads to strong families.

Chapter 16

IV. FAMILY FORM AND ETHNIC IDENTITY
   A. An analysis of family diversity requires first the recognition of the commonality of family processes among families of all types.
   B. Strengths of single-parent families include: (1) a more efficient decision-making system; (2) more direct communication; (3) a greater sense of vitality present in the work and contribution made by the child; and (4) children developing a more egalitarian view of the roles of men and women.
   C. Ethnicity is a complicated and everchanging phenomenon: Appreciating and respecting diverse types of families may shed light on how families can adapt effectively to adverse circumstances.
   D. Family strengths associated with African-American families include: (1) an extended kinship network; (2) flexibility of roles; (3) resilient children; (4) egalitarian parental relationships; and (5) strong motivation to achieve.
   E. Latino families often live in nuclear families near other families in the extended family network.
      1. Latino culture emphasizes the family as a basic source of emotional support, especially for children.
      2. In Puerto Rican families, the role of mother is central and is expressed in the term marianismo.
      3. Mexican-American families tend to emphasize the needs of the family above those of the individual: A child's padrinos (called compadres by the parents) are an important part of the family support system.
      4. Family strengths associated with Latino families include: being family centered; strong ethnic identity; high family flexibility; a supportive network of kin; equalitarian decision making; and family cohesion.
   F. Responsibilities to aged parents and to close relatives are fundamental to the family institution of Asian-American families.
      1. In Chinese-American families, the concept of hsiao (filial piety) involves a series of obligations of child to a parent.
      2. Strengths of Japanese-American families include: (1) close family ties indicated by strong feelings of loyalty to family; (2) low divorce rates; and (3) a complex system of values and techniques of social control.
      3. Due to patterns of immigration and disruption of family relations, Vietnamese-Americans have developed variations in their traditional extended family household and kin systems.
      4. Strengths of Asian-American families include: filial piety; family as a cohesive unit; value of education; feelings of loyalty; and extended family support.
   G. Native Americans are a diverse group.
      1. In general, Native American families see human life as being in harmony with nature.
      2. Relations with kin are often characterized by residential closeness, obligatory mutual aid, active participation in life cycle events, and the presence of central figures around whom family ceremonies revolve.

3. A special role for the elderly has historically been recognized as a strength in Native American families.
4. Strengths of Native American families include: extended family network; value placed on cooperation and groups, respect for the elderly; tribal support system; and preservation of culture.

V. KIN AND COMMUNITY
   A. Whether we are married or single, we have relationship needs.
      1. We need to nurture others by caring for a partner, children, or other intimates both physically and emotionally.
      2. Social integration involves being actively involved in some form of community, through knowing others who share our interests and participating in community or school projects.
      3. The knowledge that assistance from others is available keeps us from feeling anxious and vulnerable.
      4. We need intimacy with people who will listen to us and care about us.
      5. We need reassurance as to our skills as persons, workers, parents, and partners to maintain self-esteem.
   B. Few aspects of family life exist to which relatives do not make a significant contribution: Even when extended families are separated geographically, they continue to provide emotional support.
   C. In addition to kinship networks, many families have extensive networks of affiliated kin and friendships: The strength of kinship ties depends more on feeling than biology.
   D. Bronfenbrenner has proposed that we look at the family in an "ecological environment"; placing the developing child inside various systems, such as home or school.
      1. The well-being of the family depends not only on its own resources, but also on the support it receives from the community in which it is embedded.
      2. Despite the complexities of modern life, the families that love, shelter, and teach us remain America's greatest national resource: They deserve to be nurtured, strengthened, and protected.

VI. READINGS AND FEATURES
   A. In *Perspective: Tradition and Ritual in Modern American Life*, cultural traditions are examined.
      1. Our lives are measured with life rituals, mainly surrounding birth, the transition to adulthood, marriage, and death.
      2. Besides giving people a way to express their unity with nature, ritual celebrations also bring them together in a common purpose and strengthen their bonds to one another and to the community.

3. Holidays and rituals currently practiced in the United States are numerous.
   a. Yuan Tan is the Chinese New Year, typically between January 10 and February 19, and celebrated for fourteen days.
   b. Tet is the Vietnamese New Year, and is considered portentous, so families carefully monitor their words and actions to insure a year of harmony and good fortune.
   c. Holi is the North Indian festival which is characterized by noise and activity and referred to as the Festival of Colors.
   d. Pesach, the Feast of the Passover, is a Jewish festival in the spring commemorating the flight of the Jews from Egypt under the leadership of Moses.
   e. Hana Matsuri, on April 8, is celebrated by the Japanese as the birthday of the Buddha.
   f. Juneteenth is celebrated by many African Americans as the emancipation of the slaves.
   g. Ramadan is the Islamic (lunar) month-long fast where no foods are eaten by the faithful between daybreak and dusk, with a feast on the twenty-ninth night.
   h. Christmas is celebrated on December 25 by Christians to celebrate the birth of Jesus; it frequently involves feasts, the exchange of gifts, decorations, music and pageantry.
   i. Kwaanza is an African-American holiday celebrated from the day after Christmas until New Year's Day.
   j. La Quinceanera introduces young Latina girls to society on their fifteenth birthday with a Catholic mass, a feast, and a big party.
   k. Powwows are large gatherings of various Native-American groups involving demonstrations and contests of singing, dancing, and drumming.

B. The *Perspective: Cohesion and Adaptability in Stepfamilies* highlights the work of Cynthia Pill regarding stepfamilies.
   1. Healthy stepfamilies may differ from healthy intact families by being less cohesive and more adaptable.
   2. Although stepfamilies are generally less cohesive than intact families, they tend to become more cohesive over time.
   3. In creating family stories, stepfamilies begin to see themselves as uniquely different from all other families.

C. The *Family Strengths Inventory* was developed to identify areas of strengths and weaknesses in family members.

D. The *Perspective: Welfare Reform: Hardship or Hope?* discusses the welfare reform package, signed by President Clinton in 1996, as an attempt to wean families and individuals off the welfare system and help them to become more productive citizens.

SELF QUIZZES

How well do you know this material? Test your understanding of the reading assignments by answering the following sample questions.

PART I - Multiple Choice: Choose the most correct response.

_a_ 1. Which of the following statements is not true regarding marital strengths?
   a. Marital skills usually develop before family skills.
   b. Many traits of healthy marriages are also found in healthy families.
   c. Marriage may be seen as a forum for negotiating the balance between the desire for intimacy and the need to maintain a separate identity.
   d. Relationships of couples with children generally have greater stability than those of childless couples.
   e. all of the above

_c_ 2. Therapist-researcher David Mace cites the essential aspects of successful marriage as
   a. affirmation, respect, and trust.
   b. role modeling, affirmation, and play.
   c. commitment, communication, and creative use of conflict.
   d. responsibility, self-respect, and a moral code.
   e. tradition, ritual, and a sense of family history.

_c_ 3. The services provided by families
   a. are usually appreciated by society.
   b. could easily be replaced within communities.
   c. are basic to human existence.
   d. have little impact outside the family itself.
   e. are often highlighted in the media.

_c_ 4. Commitment involves all but which one of the following?
   a. helping other family members actualize their potential
   b. sacrifice
   c. obligation and duty
   d. the participation of family members in a world view that encompasses more than self-centered interest
   e. balancing self and family

_e_ 5. Positive parental role-modeling
   a. is a crucial factor in the development of qualities that ensure personal psychological health and growth of children.
   b. is important because a loving relationship between parents seems to breed security in the children.
   c. is important because secure children are more able to take risks and reach out to others.
   d. is important because secure children are more able to become independent and develop a good self-image.
   e. all of the above

Chapter 16

___e___ 6. Good communication
   a. is an art that is well developed in strong families.
   b. is easily achieved in most families.
   c. facilitates other family strengths.
   d. is direct.
   e. all but b

___d___ 7. Which of the following is not true of responsibility within the family?
   a. Responsibility stems from a sense of self-respect and an appreciation of the interdependence of the family members upon each other.
   b. It involves understanding how much difference our own acts make in the lives of other people.
   c. Delegating responsibility includes participation in household chores and family decisions.
   d. Forcing children to accept more responsibility than they want is usually harmful to them.
   e. Allowing children to make their own mistakes and to face the consequences of their actions is an important lesson in developing responsibility.

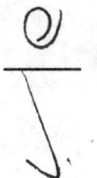

___e___ 8. Coping strategies healthy families possess include
   a. spiritual resources.
   b. seeking outside help, if needed.
   c. adaptability.
   d. a sense of unity.
   e. all of the above

___c___ 9. All but which one of the following are true regarding strong families and time?
   a. Healthy families give play and leisure time high priority.
   b. One of the first signs of family difficulties is lack of family time together.
   c. As children arrive, having time to be together becomes less important.
   d. Children learn they are valued and appreciated when family members spend time with them.
   e. Quality time need not be spent on lengthy or expensive vacations.

___b___ 10. Physical, emotional, and psychological factors which define space and separate family members from one another are known as
   a. coalitions.
   b. boundaries.
   c. bondedness.
   d. enmeshment factors.
   e. disengagement factors.

Marriage and Family Strengths

___b___ 11. Which of the following is not a dimension critical to family adaptability?
   a. assertiveness
   b. boundaries
   c. discipline
   d. negotiation
   e. roles

___a___ 12. African-American families
   a. maintain a pedi-focal orientation.
   b. interact infrequently with extended kin.
   c. model traditional roles.
   d. tend to devalue education.
   e. ignore the realities of being black in America.

___e___ 13. In Chinese-American families, the concept of hsiao involves
   a. parental obligations to children.
   b. child obligations to parents.
   c. filial piety.
   d. differing obligations according to birth order.
   e. all but a

___b___ 14. Major family strengths of Native Americans include all but which of the following?
   a. value placed on cooperation
   b. equalitarian decision making
   c. preservation of culture
   d. respect for the elderly
   e. extended family network

___c___ 15. Which of the following is not identified by Robert Weiss as needs that can be met only in relationships?
   a. nurturing others
   b. intimacy
   c. support systems
   d. reassurance
   e. social integration

___d___ 16. Which of the following ideas is not congruent with the concept of family ecology.
   a. The family exists in the environment of community and is influenced by surrounding social change.
   b. Policies supportive of families are crucial to its survival.
   c. The "ecological environment" can be conceptualized as a set of nested Russian dolls.
   d. The well-being of the family depends entirely upon its own resources.
   e. The interplay of many systems profoundly affects the child in their midst.

Chapter 16

PART II - True/False

_T_ 1. Interpersonal competence is necessary to negotiate the balance between the desire for intimacy and the need to maintain one's separate identity.

_T_ 2. Childfree couples generally have more time for each other and significantly less psychological, economic, and physical stress.

_T_ 3. Research indicates the strongest predictor of marital success may be the effectiveness of communication experienced by the couple before marriage.

_F_ 4. When spouses who love each other have sexual liaisons outside the marriage, it does not necessarily mean the marriage is in trouble. (Assoc or relation)

_T_ 5. Family quality varies over the family life cycle.

_T_ 6. All families constantly grow and change.

_F_ 7. Families that hamper (caused difficulty) the expression of their children's beliefs tend to send into society children who respect others.

_T_ 8. Most youth with conduct disorders come from family situations characterized by poor communication.

_F_ 9. Spiritual wellness is the same as religiosity.

_F_ 10. Focusing on family history and family traditions is relatively unimportant in strengthening the family.

_F_ 11. Healthy families are problem free.

_F_ 12. To secure outside help or intervention with internal family problems is an admission of weakness and failure.

_T_ 13. Spending time together is necessary to develop communication and to build cohesion.

_T_ 14. Leadership in strong families moves from one person to another according to the area of family activity.

_F_ 15. Coalitions (union) usually decrease family cohesiveness. (act or sticking together)

_F_ 16. Latino cultures clearly define the difference between relatives and friends.

_T_ 17. The well-being of the family depends on the support of the larger community.

_F_ 18. Healthy stepfamilies are less adaptable than healthy intact families.

266

Marriage and Family Strengths

DISCUSS BRIEFLY

Below is a list of qualities of healthy families found within Chapter Sixteen. After each quality, define the term or concept and discuss how it contributes to family strength.

1. Interpersonal competence:

2. Commitment:

3. Communication:

4. Creative use of conflict:

5. Affirmation:

6. Respect for self and others:

7. Trust:

8. Role-modeling:

9. Responsibility:

10. Morality:

(continued on next page)

Chapter 16

(continued from previous page)

11. Spiritual orientation:

12. Traditions and rituals:

13. A sense of family history:

14. Willingness to seek help in a crisis:

15. Play and leisure time:

16. Cohesiveness:

17. Adaptability:

18. Ethnic Identity

19. Kinship ties:

20. Affiliated kin:

21. Friendship:

22. Community ties:

SELF-DISCOVERY

Select three of the previous traits or qualities that you would like to see enhanced within your family. Why did you select these three? What particular techniques can you personally use to develop or enhance these traits?

1.

2.

3.

MINI-ASSIGNMENT

Select one of the three traits above and attempt to actively increase or improve it within your own family. Which did you select? What did you do specifically within your family to enhance this quality?

What were the reactions of your family to this? Were they aware of your intentions?

Would you continue this action or attempt it again? Was this a positive experience for you?

Chapter 16

SELF-REFLECTION

What were some of the most important aspects of marriage, relationships and/or the family which you learned from this text?

What were the specific things from this book which you were able to immediately put to use?

How has/will the information gained from the text impact your present relationships or future relationships?

KEY TO SELF QUIZZES

Multiple Choice   **True/False**

1. a   10. b   1. T   10. F
2. c   11. b   2. T   11. F
3. c   12. a   3. T   12. F
4. c   13. e   4. F   13. T
5. e   14. b   5. T   14. T
6. e   15. c   6. T   15. F
7. d   16. d   7. F   16. F
8. e   8. T   17. T
9. c   9. F   18. F

SUGGESTED READINGS

For related readings, see 589 in the text.